Atmospheric Pressure Plasmas at Interfaces

Online at: https://doi.org/10.1088/978-0-7503-6254-2

IOP Series in Plasma Physics

Series Editor

Richard Dendy

Culham Centre for Fusion Energy and the University of Warwick, UK

About the series

The IOP Plasma Physics ebook series aims at comprehensive coverage of the physics and applications of natural and laboratory plasmas, across all temperature regimes. Books in the series range from graduate and upper-level undergraduate textbooks, research monographs and reviews.

The conceptual areas of plasma physics addressed in the series include:

- Equilibrium, stability, and control.
- Waves: fundamental properties, emission, and absorption.
- Nonlinear phenomena and turbulence.
- Transport theory and phenomenology.
- Laser–plasma interactions.
- Non-thermal and suprathermal particle populations.
- Beams and non-neutral plasmas.
- High energy density physics.
- Plasma–solid interactions, dusty, complex, and non-ideal plasmas.
- Diagnostic measurements and techniques for data analysis.

The fields of application include:

- Nuclear fusion through magnetic and inertial confinement.
- Solar–terrestrial and astrophysical plasma environments and phenomena.
- Advanced radiation sources.
- Materials processing and functionalisation.
- Propulsion, combustion, and bulk materials management.
- Interaction of plasma with living matter and liquids.
- Biological, medical, and environmental systems.
- Low-temperature plasmas, glow discharges, and vacuum arcs.
- Plasma chemistry and reaction mechanisms.
- Plasma production by novel means.

A list of recent titles published in this series can be found here: https://iopscience.iop.org/bookListInfo/iop-plasma-physics-series.

Atmospheric Pressure Plasmas at Interfaces

Katharina Stapelmann
Department of Nuclear Engineering, North Carolina State University, Raleigh, NC, USA

IOP Publishing, Bristol, UK

ISBN 978-0-7503-6254-2 (ebook)
ISBN 978-0-7503-6252-8 (print)
ISBN 978-0-7503-6255-9 (myPrint)
ISBN 978-0-7503-6253-5 (mobi)

DOI 10.1088/978-0-7503-6254-2

Version: 20251001

IOP ebooks

British Library Cataloguing-in-Publication Data: A catalogue record for this book is available from the British Library.

Published by IOP Publishing, wholly owned by The Institute of Physics, London

IOP Publishing, No.2 The Distillery, Glassfields, Avon Street, Bristol, BS2 0GR, UK

US Office: IOP Publishing, Inc., 190 North Independence Mall West, Suite 601, Philadelphia, PA 19106, USA

Contents

Preface

Many low temperature plasma applications at atmospheric pressure involve interactions with solid or liquid surfaces. Reactive species produced by the plasma interact with a target which could be skin or cell culture for plasma medicine applications or nitrogen fixation in a liquid solution for plasma agriculture applications. In many cases, the plasma is interfacing a liquid layer. While there is a wealth of literature for low-pressure plasma surface interactions, the literature on atmospheric pressure plasmas interfacing solid or liquid surfaces is scarce. This eBook aims to introduce physics and chemistry at interfaces (solid and liquid) and in particular plasmas at interfaces. The eBook starts with a general introduction to surfaces and interfaces, as well as plasma–surface interactions. Plasma–solid and plasma–liquid interactions are the main focus of the book, followed by a chapter devoted to examples of applications including plasma–bio interactions and pollutant removal. With a general introduction to interfaces, this eBook seeks to bridge a gap between interface sciences, chemistry, physics, and plasma physics/engineering. This book is meant to enable researchers to find an introduction to various topics related and fundamental for plasma–liquid interactions and create thus the scientific basis for various applications involving plasmas at interfaces.

Author biography

Katharina Stapelmann

Katharina Stapelmann has dedicated her career to unlocking the potential of plasmas in life sciences. Now an Associate Professor of nuclear engineering at North Carolina State University, she leads research that bridges fundamental plasma physics with real-world applications in medicine and agriculture.

Her path began in Germany, where she earned her PhD in electrical engineering from Ruhr University Bochum in 2013. After serving there as a postdoctoral researcher and assistant professor, she moved to the United States to continue her work at NC State in 2017. Recognized internationally, she is a Mercator Fellow of the German Research Foundation (DFG), trustee of the German National Center for Plasma Medicine, and editor-in-chief for *Plasma Processes and Polymers*. Her achievements have been honored with the Goodnight Early Career Innovators Award and the Early Career Award in Plasma Medicine.

Driven by curiosity and the belief that plasmas can help solve pressing challenges, she continues to push the boundaries of plasma science at the interface with life.

Symbols

Plasma and discharge parameters

E	Electric field strength (V m^{-1})
V	Applied voltage (V)
J	Current density (A m^{-2})
n_e	Electron density (m^{-3})
n_i	Ion density (m^{-3})
μ	Mobility (m^2 V^{-1} s^{-1})
σ	Conductivity (S m^{-1} or μS cm^{-1} in liquids)
T_g	Gas temperature
Q	Total charge
γ	Secondary electron emission coefficient
α	Ionization coefficient
$\sigma(v)$	Cross-section
R	Reaction rate
$f(E)$	Electron energy distribution function
$f(v)$	Electron velocity distribution function
ρ_e	Total electric charge density
$q = \pm e$	Elementary charge
λ_D	Debye length
ν_m	Momentum transfer collision frequency

Transport and kinetics

D	Diffusion coefficient (m^2 s^{-1})
k	Reaction rate constant (varies with order)
Γ	Flux across interface (mol m^{-2} s^{-1})
H_i^{cc}	Henry's law constant (concentration–concentration form)
c_g	Gas-phase concentration (mol m^{-3})
c_l	Liquid-phase concentration (mol m^{-3})
Δx	Diffusion boundary layer thickness (m)

Surface, liquid, and interface

θ	Surface coverage (–)
Δp	Laplace pressure (Pa)
$\Delta\phi$	Potential drop across sheath or double layer (V)
R	Droplet radius or radius of curvature (m)
γ	Surface tension (N m^{-1})
γ_s	Surface energy
η	Viscosity (Pa s)
U	Voltage
E_a	Chemical reaction activation energy
E_s	Excess energy at surface
Σ_A	Molecular surface area
σ_{stress}	Surface stress
ε_{total}	Surface strain
κ	Capillary length
ρ	Density (liquid)
V_m	Molar volume of a liquid

μ_i Chemical potential
Γ_i Interfacial excess or surface concentration
s Sticking coefficient
β Surface loss probability
r Reflection coefficient

Photons and energy

$h\nu$ Photon energy (J or eV)
λ Wavelength (m or nm)

Constants

R Gas constant
k_B Boltzmann constant
ε Dielectric constant/permittivity (–)
ε_0 Vacuum permittivity (F m^{-1})
ε_{eff} Effective permittivity

Miscellaneous

t Time (s)
τ Characteristic time scale (s)
τ_M Maxwell relaxation time
W Work

Abbreviations

AC	Alternating current
ALD	Atomic layer deposition
APPJ	Atmospheric pressure plasma jet
DBD	Dielectric barrier discharge
DC	Direct current
EEDF	Electron energy distribution function
EHD	Electrohydrodynamic
IEDF	Ion energy distribution function
KH	Kelvin–Helmholtz (instability)
PECVD	Plasma enhanced chemical vapor deposition
RF	Radio frequency
RT	Rayleigh–Taylor (instability)
UV	Ultraviolet
VUV	Vacuum ultraviolet

Chapter 1

Introduction

Atmospheric-pressure plasmas at interfaces are powerful because fields, charges, photons, and chemically active species all couple to surfaces and liquids whose properties evolve in time. This eBook builds a practical framework for understanding and using that coupling across materials, geometries, and timescales. It assumes familiarity with basic plasma concepts and focuses on what matters at (and within a few hundred nanometers of) the interface: adsorption and solvation, transport and renewal, sheath–double-layer coupling and Faradaic pathways, photon/metastable effects, and plasma-induced fluid motion.

How the book is organized. Chapter 2 introduces the physics and chemistry of solid and liquid interfaces (surface energy, Laplace/Kelvin relations, Gibbs adsorption, electric double layers), establishing the thermodynamic and electrostatic language used later. Chapter 3 surveys atmospheric-pressure discharges and the interface-relevant 'ingredients' (gas-phase chemistry, charged species and sheaths, UV/VUV and metastables), followed by surface kinetics for low- and high-energy particles. Chapters 4 and 5 are the core: plasma–solid and plasma–liquid interactions, each treating neutrals, charged species (including double layers, solvated electrons, and Faradaic steps), photons/metastables, and the consequences of direct versus indirect sources. Chapter 6 applies the framework to two high-impact areas: plasma–bio systems and water treatment, where design choices, transport, and timescales decide outcomes.

What this means for practice. In plasma–liquid systems, electrical coupling, interfacial fields, double-layer structure, solvated electrons, and Faradaic pathways tie plasma dynamics to electrochemistry (section 5.3), while photons and metastables act as predominantly interfacial drivers with short penetration depths (section 5.4). Plasma-induced flows (Maxwell stress, ion momentum, Marangoni, gas-jet shear) reshape the interface and feed back on transport (section 5.5). Use these chapters as a map: choose modeling boundaries (Henry's law versus film theory versus reaction–diffusion; add convection when jets/flows are present), match

doi:10.1088/978-0-7503-6254-2ch1

agents to targets (oxidants versus interfacial reduction), and engineer renewal/ residence time at the interface before tweaking chemistry.

Caution: Some symbols used within the eBook are the same but with different meanings. This is primarily because the eBook brings together information from different fields. Instead of coming up with new variables, the author decided to have the field-specific terminology preserved throughout with the hope that the context makes it clear which variable is actually used.

IOP Publishing

Atmospheric Pressure Plasmas at Interfaces

Katharina Stapelmann

Chapter 2

Introduction to surfaces and interfaces

The main focus of the book is the interaction of plasmas with interfaces. It is assumed that the reader is familiar with the basic concepts of a plasma. However, the typical plasma researcher might not be as familiar with surfaces and interfaces. A brief introduction to the physics and chemistry of interfaces is thus provided here. For the more interested reader, recommendations for further reading are made at the end of the chapter.

2.1 Surfaces and interfaces

At-a-glance
Surfaces and interfaces are distinct from bulk materials because particles at the boundary lack full coordination. This gives rise to unique physical and chemical properties that dominate processes at the phase boundary.

According to the *Encyclopedia Britannica*, a surface is defined as 'Surface, in chemistry and physics, the outermost layer of a material or substance. Because the particles (atoms or molecules) on the surface have nearest neighbours beside and below but not above, the physical and chemical properties of a surface differ from those of the bulk material; surface chemistry is thus a branch of physical chemistry' [1], while an interface is defined as: 'Interface, surface separating two phases of matter, each of which may be solid, liquid, or gaseous. An interface is not a geometric surface but a thin layer that has properties differing from those of the bulk material on either side of the interface. A common interface is that between a body of water and the air, which exhibits such properties as surface tension, by which the interface acts somewhat like a stretched elastic membrane. Interfacial effects, or processes that occur at interfaces, include the evaporation of liquids, the action of

doi:10.1088/978-0-7503-6254-2ch2

detergents and chemical catalysts, and the adsorption of gases on metals' [2]. The terms surface and interface are often used synonymously. The term interface is preferred to indicate the boundary between phases. For example, it is a solid surface but a solid–gas interface.

The physical and chemical properties of a surface are different than those of the bulk material. Molecules at a surface are arranged differently as they do not have a neighbor above. In general, this is the case for liquid and solid surfaces. However, at a gas–liquid interface, there is no barrier preventing molecules from leaving or entering the surface [3]. A solid surface is compact and tightly packed with molecules or atoms. It remains in a fixed shape and volume (usually), and molecules are not mobile; they only vibrate in place. Liquid surfaces on the other hand have loosely packed molecules and they can flow and move past each other. In the following, solid and liquid surfaces will be introduced in more detail.

2.2 Solid surfaces

At-a-glance

Solid surfaces represent a termination of the bulk lattice, leading to distinct properties such as reconstruction, oxidation, and sensitivity to probes. For many applications, including plasma interactions, the surface cannot be neglected. The broken symmetry of surfaces causes atomic rearrangements, modified electronic states, and distinct energetics. These determine surface free energy, tension, and stress, which in turn govern stability and interaction with external species.

Solid surfaces mark the boundary between a material and its surrounding environment. At the atomic level, a surface represents a termination of the periodic lattice that defines the solid's bulk. This disruption in symmetry gives rise to unique physical and chemical properties that differ significantly from those of the interior. Understanding how solid surfaces behave is essential for interpreting how they respond to ions, photons, or reactive species from a plasma. Before plasma–surface interactions can be discussed meaningfully, it is important to explore the fundamental structure, energetics, and behavior of solid surfaces in the absence of plasma. Solid surfaces are formed by a relatively small number of atoms and molecules ($\approx 10^{15}$ cm^{-2}). In classical solid-state theory or the atomic physics of condensed matter, surfaces are typically neglected by assuming infinitely extended solids [4]. For the derivation of macroscopic properties of a solid that contains about 10^{23} atoms in 1 cm^{-3}, e.g., elasticity, specific heat, or conductance, the surface is, indeed, negligible. Even for the theoretical description of spectroscopic experiments the simplified infinitely extended solid theory is still valid as long as the probes can penetrate deep enough into the solid (e.g., x-rays, fast electrons). If the probes only penetrate a few angstroms, or if surface interaction due to photoemission or other excitation mechanisms (like it is often the case in plasma-facing materials) becomes important, the infinitely extended solid theory is invalid as the few molecules or atoms that make up the surface dominate the response and cannot be neglected anymore. One difference between the bulk and the surface material is that most solids are not crystalline at the surface, even if the bulk is crystalline or polycrystalline. Many surfaces show chemical changes as well, for example they get oxidized under ambient conditions, e.g., aluminum (Al), which forms a hard oxide layer (Al_2O_3) immediately in contact with air [3]. Real surfaces in ambient conditions at atmospheric pressure are thus a difficult system to study and far away from being well-defined. In the field of surface and interface science, experiments are often conducted under ultra-high vacuum conditions with fresh, clean surfaces [4] to gain understanding of surface processes. The simplest (yet still complex) interface to study is that between a crystalline surface and vacuum. Butt *et al* point out: 'Since many natural surfaces are amorphous, it may seem somewhat academic to study crystalline surfaces' [3], followed by a list of reasons why it is still a good idea to do so; (a) it allows for comparison of experiments on different samples of the same material as the structure of the crystalline surfaces is well-defined, (b) the periodic structure of the surfaces aids in the theoretical description allowing the use of diffraction methods to analyze the surfaces, and (c) the semiconductor industry depends on the defined production of crystalline surfaces.

2.2.1 Structure and properties of solid surfaces

In crystalline solids, atoms are arranged in a periodic lattice that extends in three dimensions. At the surface, the atoms experience a different environment as they are no longer symmetrically surrounded by neighbors. This lack of coordination leads to atomic rearrangements known as surface relaxation (a small shift in atomic positions) and surface reconstruction (a change in surface periodicity), especially on low-index surfaces such as (100), (110), or (111), to minimize the less energetic favorable dangling bonds. The surface relaxation and reconstruction is a temperature-dependent process.

The broken symmetry at a surface not only affects the atomic arrangement but also modifies the electronic structure. Electrons at the surface occupy states that are distinct from those in the bulk. These surface states arise due to the discontinuity in the potential at the interface and can dominate the chemical and physical interactions at the surface. In semiconductors and insulators, the surface may also exhibit band bending, a shift in energy bands near the surface due to charge redistribution, especially in the presence of adsorbates or surface defects. For example, oxygen adsorbed on a silicon surface can lead to significant band bending, affecting the surface potential and electronic transport. The electronic properties of a surface govern its interaction with charged plasma species, its ability to emit or trap electrons, and its reactivity with radicals or photons. For example, electron emission from the surface can be a key process that can impact the plasma behavior, as we will explore in section 3.2.2.

Creating a surface requires breaking atomic bonds, which costs energy. This leads to the concept of surface free energy, usually expressed in J m^{-2}. An excess energy E_S is associated with each surface atom and dividing that by the atomic or molecular surface area Σ_A yields the surface energy γ_S. For any reversible process, the work done to create a new surface area is a product of surface energy γ_S, a generalized force, and the increase in surface area dA, a generalized displacement: $dW = \gamma_S \, dA$. This equals the increase in excess energy E_S and number of surface atoms N:

$$dW = \gamma_S \, dA = d(E_S \, N) = E_S \frac{\partial N}{\partial A} dA + N \frac{\partial E_S}{\partial A} dA \tag{2.1}$$

or:

$$\gamma_S = E_S \frac{\partial N}{\partial A} + N \frac{\partial E_S}{\partial A} \tag{2.2}$$

The two parts of equation (2.2) can be interpreted as plastic (reversible change in surface area by bringing in new atoms to the surface without changing their relative distance) and elastic (N remains constant) contributions to surface energy. For a plastic change in surface area: $dA = \Sigma_A dN$. The excess energy of an atom E_S does not depend on the number of atoms N, therefore:

$$\left(\frac{\partial E_S}{\partial A}\right)_{\text{plastic}} = \frac{\partial E_S}{\partial (N\Sigma_A)} = \frac{1}{\Sigma_A}\frac{\partial E_S}{\partial N} = 0 \tag{2.3}$$

For the plastic process, equation (2.2) becomes:

$$\gamma_{S,\text{plastic}} = \left(E_S \frac{\partial N}{\partial A}\right)_{\text{plastic}} = E_S \frac{\partial A/\Sigma_A}{\partial A} = \frac{E_S}{\Sigma_A} \equiv \gamma \tag{2.4}$$

The change in surface energy in case of a plastic change in surface area is similar to that of a liquid and is thus called surface tension γ.

In an elastic change of the surface area, N is constant: $dA = N d\Sigma_A$ and $\partial N/\partial A = 0$ and equation (2.2) becomes:

$$\gamma_{S,\text{elastic}} = \left(N \frac{\partial E_S}{\partial A}\right)_{\text{elastic}} = \frac{N}{N}\frac{\partial E_S}{\partial A} = \frac{\partial E_S}{\partial \Sigma_A} \tag{2.5}$$

Combining the equations:

$$\gamma_{S,\text{elastic}} = \frac{\partial(\gamma \Sigma_A)}{\partial \Sigma_A} = \gamma + \frac{\partial \gamma}{\partial \Sigma_A} \tag{2.6}$$

It is convenient now to introduce the surface strain. It is defined by the change in surface area divided by the whole area: $d\varepsilon_{\text{tot}} = dA/A = d\varepsilon_{\text{plastic}} + d\varepsilon_{\text{elastic}}$. For an elastic change, only $\varepsilon_{\text{elastic}}$ needs to be taken into account: $d\varepsilon_{\text{elastic}} = dA/A = d\Sigma_A/\Sigma_A$. Inserting yields:

$$\gamma_{S,\text{elastic}} = \gamma + A \frac{\partial \gamma}{\partial A} = \gamma + \frac{\partial \gamma}{\partial \varepsilon_{\text{elastic}}} \equiv \sigma_{\text{stress}} \tag{2.7}$$

The change in elastic surface energy is therefore a sum of the surface tension and its derivative with respect to the elastic surface strain, which defines the surface stress σ_{stress}.

2.3 Liquid surfaces

At-a-glance

Liquid interfaces are dynamic and fluctuate at molecular scales. They exhibit gradients in density and orientation, giving rise to surface tension and governing how molecules evaporate, condense, or exchange across the interface. Thermodynamic relations such as Laplace pressure, Kelvin equation, and Gibbs adsorption describe the equilibrium shape, vapor pressure, and surfactant effects at liquid surfaces. Local surface tension gradients can drive flows (Marangoni effect). Charged liquid surfaces form electric double layers, described by the Poisson–Boltzmann framework. Screening by ions defines the Debye length, capacitance, and surface potential, which are central to plasma–liquid electrochemical behavior.

In contrast to solids, the surface of a liquid is a dynamic, deformable, and molecularly heterogeneous region. Understanding the properties of liquid interfaces is essential for describing plasma–liquid interactions, particularly at atmospheric pressure, where effects such as evaporation, convective transport, and chemical reactivity become significant. Unlike the rigid and structured nature of solid surfaces, liquid surfaces lack long-range order and are subject to thermal fluctuations that constantly reshape their microscopic topography. On a microscopic level, the liquid surface does not have a sharp boundary in the direction of its normal. Hence, the liquid interface stretches over some thickness with a density gradient decreasing from the bulk liquid to that of its vapor [3]. Liquid surfaces are in constant movement. Water molecules evaporate from the liquid into the vapor phase and water molecules enter the liquid from the gas phase. In addition, surface-near water molecules diffuse into the bulk liquid and bulk molecules diffuse to the surface. In equilibrium at room temperature, the average time a water molecule spends at the surface is on sub-microsecond scale ($\approx 0.1\ \mu s$) [3]. The interfacial region typically spans a few nanometers in depth and is characterized by gradients in density, dielectric properties, and hydrogen bonding (in the case of polar liquids like water). Techniques such as x-ray reflectivity and vibrational sum-frequency generation (SFG) spectroscopy have shown that water surfaces, for instance, exhibit a degree of molecular orientation, with O–H bonds preferentially pointing into the vapor phase [5].

2.3.1 Structure and properties of liquid surfaces

Several thermodynamic concepts govern the structure and properties of liquid surfaces. Figure 2.1 gives a conceptual overview of the concepts briefly introduced in this section.

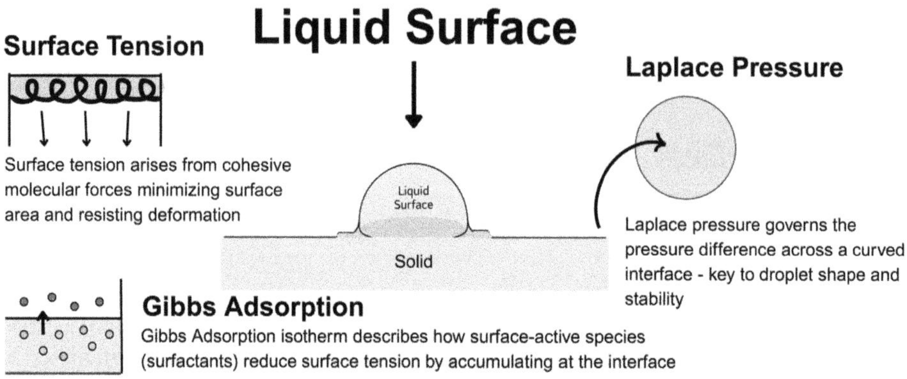

Figure 2.1. Conceptual overview of the thermodynamic concepts governing the structure and properties of a liquid surface described in this section.

Table 2.1. Surface tensions of some liquids relevant for plasma–liquid interactions. PFOA: perfluorooctanoic acid. PFOS: perfluorooctane sulfonate. Both belong to the larger group of PFAS (per- and polyfluoroalkyl substances). PBS: Phosphate buffered saline, a commonly used buffer for plasma treatments of liquids. DMEM: Dulbecco's Modified Eagle Medium, widely used cell culture medium.

Substance	T/°C	γ/mN m^{-1}	Reference
Water	25	71.99	[3]
PFOA (100 mg l^{-1})	22	≈ 65	[6]
PFOA (1000 mg l^{-1})	22	≈ 45	[6]
PFOS (100 mg l^{-1})	22	≈ 55	[6]
PFOS (1000 mg l^{-1})	22	≈ 35	[6]
PBS	20	≈ 60	[7]
DMEM	20	≈ 70	[8]

At the molecular level, the surface of a liquid is defined as the interface where cohesive forces within the liquid are no longer fully balanced by neighboring molecules. Molecules at or near the surface experience an anisotropic environment, with fewer neighboring molecules on the vapor side. This leads to a net inward force, giving rise to surface tension γ, a key thermodynamic parameter that governs interfacial phenomena. It can also be viewed as the energy required to bring molecules from the bulk liquid to the surface. The surface tension of liquids decreases with temperature T [3]:

$$\gamma(T) = \gamma(T_0) + \frac{\partial\gamma}{\partial T}|_{T=T_0}(T - T_0) \tag{2.8}$$

where the coefficient $\frac{\partial\gamma}{\partial T}$ is negative and relates to the surface entropy. Table 2.1 gives an overview of surface tension values of some liquids relevant for plasma–liquid interactions. Water has the highest surface tension with 71.99 mN m^{-1}. Perfluorooctanoic acid (PFOA) and perfluorooctane sulfonate (PFOS) both belong to the larger group of PFAS (per- and polyfluoroalkyl substances), so called 'forever molecules', serving as one of the application examples in chapter 6. PFAS are surfactants and reduce the surface tension of the solution with increasing concentration. Phosphate buffered saline (PBS) is a commonly used buffer for plasma treatments of liquids and has a lower surface tension than water. Dulbecco's Modified Eagle Medium (DMEM) is a widely used cell culture medium with a similar surface tension as water. As we will explore in section 5.5.2, the balance between surface tension and electric field determines interfacial deformations, oscillations, and instabilities.

Due to surface tension, there is a capillary pressure difference sustained between two static fluids, such as water and air. The Young–Laplace equation relates the difference in pressure between the two phases ΔP, also called Laplace pressure, to the curvature of the surface with the two principal radii of curvature R_1 and R_2, in the absence of any external forces:

$$\Delta P = \gamma\left(\frac{1}{R_1} + \frac{1}{R_2}\right) \tag{2.9}$$

With this equation, the equilibrium shape of a liquid surface can be determined, or the pressure difference when the radii are known. For example, a spherical bubble with a diameter of 2 mm ($R_1 = R_2 = R = 1$ mm) experiences:

$$\Delta P = \frac{2\gamma}{R} = \frac{2 \times 71.99 \text{ mN m}^{-1}}{1 \text{ mm}} = 143.98 \text{ N m}^{-2} \approx 144 \text{ Pa} \tag{2.10}$$

A smaller bubble of 20 nm diameter would accordingly experience a much larger pressure difference: $\Delta P = \frac{2 \times 71.99 \text{ mN m}^{-1}}{10^{-8} \text{ m}} = 1.44 \times 10^7 \text{ N m}^{-2}$. The pressure inside the bubble is 1.44×10^7 Pa larger than the outside pressure.

With the absence of external fields (e.g., gravity), the Laplace pressure is the same everywhere in the liquid: $\Delta P = \text{constant}$. There would be a flow to regions of low pressure otherwise. It follows that for constant ΔP, the surface of the liquid has the same curvature everywhere.

When does gravity, or other external forces have to be taken into account? The hydrostatic pressure caused by gravitation needs to be taken into account when the structure is large, determined by the characteristic length scale κ:

$$\kappa = \sqrt{\frac{\gamma}{g\rho}} \tag{2.11}$$

where κ is also called capillary length or capillary constant. g is the acceleration of free fall and ρ is the density of the liquid. If the curvature of the liquid structure is much smaller than κ, gravitation can be neglected. Otherwise, the Young–Laplace equation (2.9) needs to include a term to include the hydrostatic pressure caused by gravitation:

$$\Delta P = \gamma\left(\frac{1}{R_1} + \frac{1}{R_2}\right) + \rho g h \tag{2.12}$$

where h is the height coordinate.

Another fundamental equation in surface sciences based on thermodynamic principles is the Kelvin equation [9]. The vapor pressure depends on the curvature of the liquid surface due to the above introduced Laplace pressure:

$$RT \ln \frac{P_0^{\mathrm{K}}}{P_0} = \gamma V_{\mathrm{m}}\left(\frac{1}{R_1} + \frac{1}{R_2}\right) \tag{2.13}$$

with R the universal gas constant, P_0^{K} the vapor pressure of the curved surface, P_0 the vapor pressure of a flat surface, and V_{m} the molar volume of the liquid. The index '0' represents thermodynamic equilibrium (in equilibrium, the curvature of the liquid surface is constant). To illustrate the Kelvin equation, we consider two distinct cases: (i) a drop (positively curved liquid surface) and (ii) a bubble (negatively curved liquid surface). In the first case, the vapor pressure of the drop is higher, with smaller drops exhibiting a larger vapor pressure than larger drops. The raised pressure makes evaporation easier, so that smaller drops evaporate more quickly. The resulting vapor then tends to condense into larger drops. For the second case, the vapor pressure of the liquid with respect to the bubble is reduced because of the negative curvature of the liquid, making it more difficult for molecules from the liquid to evaporate.

Lastly, Gibbs adsorption isotherm describes the change in surface tension upon adsorption of substances to the interface:

$$\mathrm{d}\gamma = -\sum \Gamma_i \mathrm{d}\mu_i \tag{2.14}$$

with Γ_i the interfacial excess or surface concentration, and μ_i the chemical potential. This isotherm is valid for constant temperatures. More general descriptions and the derivation can be found in [3]. The change in surface tension is visible, e.g., when detergents are added to water. Detergents like soap reduce the surface tension which makes wetting and cleaning easier. Similarly, many PFAS chemicals are surfactants and will reduce the surface tension of the liquid. A simple case to apply Gibbs adsorption isotherm is a system of two components like a solvent 1 and a solute 2:

$$\Gamma_2^{(1)} = -\frac{a}{RT} \cdot \frac{\partial \gamma}{\partial a} \tag{2.15}$$

When a solute is enriched at a surface, $\Gamma_2^{(1)} > 0$. The surface tension decreases with increasing concentration. These substances are surface active agents and called surfactants. When $\Gamma_2^{(1)} < 0$, the solute avoids the interface and surface tension will increase with increasing concentration. Adding impurities like salt to water would be an example for this.

Local variations in surface tension will cause movement at a fluid interface. This is called the Marangoni effect and will be further explored in section 5.5.1. A gradient in surface tension causes a hydrodynamic drag leading to a liquid flow.

Gradients in surface tension can be caused by a local temperature variation (cooling due to evaporation or localized heating) and local variations in composition. Temperature-induced surface tension gradients cause an instability that leads to Bénard–Marangoni convection.

2.3.2 Electrical and chemical characteristics

Water with its high permittivity of $\varepsilon \approx 80$ is a good solvent for ions, so most surfaces in water are charged. For liquids with lower permittivity, e.g., organic liquids, the charging of surfaces is negligible. Ions can adsorb to a surface or dissociate from a surface. Proteins or amino acids, for example, expose the amino group on their surface which becomes protonated and positively charged ($R\text{–}NH_2 + H^+ \rightarrow R\text{–}NH_3^+$). Oxides, on the other hand, are often negatively charged due to the dissociation of a proton from a surface hydroxyl group. The surface charges cause an electric field which in turn attracts counterions, giving rise to the formation of an electric double layer. Here, we will focus on the Gouy–Chapman model of the electric double layer (for further reading and extended models, see, e.g., [3]). The description of the electric double layer can be performed with equations that should be well-known to the plasma community; it is based on Poisson's equation and Boltzmann statistics. Solvents are treated as continuous media with a dielectric constant ε, ignoring the molecular nature of a liquid. The potential $\phi(x, y, z)$ near a charged surface can be derived using Poisson's equation:

$$\nabla^2\phi = \frac{\partial^2\phi}{\partial x^2} + \frac{\partial^2\phi}{\partial y^2} + \frac{\partial^2\phi}{\partial z^2} = -\frac{\rho_e}{\varepsilon\varepsilon_0} \tag{2.16}$$

with ρ_e the local electric charge density. The ions in the liquid can move freely and their local density is described by the Boltzmann equation:

$$n_i = n_{i,0} \cdot \exp\frac{W_i}{k_B T} \tag{2.17}$$

with W_i the electric work required to bring an ion to the potential ϕ ($W_i = \pm e\phi$, + for cation, − for anion) and $n_{i,0}$ the ion density in the bulk liquid. Inserting $\rho_e = e(n_{i,+} - n_{i,-})$ into equation (2.16) yields:

$$\nabla^2\phi = \frac{en_{i,0}}{\varepsilon\varepsilon_0}\left(\exp\frac{e\phi(x, y, z)}{k_B T} - \exp\frac{-e\phi(x, y, z)}{k_B T}\right) \tag{2.18}$$

This partial differential equation of second order, also known as the Poisson–Boltzmann equation, can only be solved analytically for simple geometries like planar surfaces. For planar surfaces, the potential cannot change in y and z, simplifying Poisson–Boltzmann equation to a one-dimensional problem:

$$\frac{d^2\phi}{dx^2} = \frac{en_{i,0}}{\varepsilon\varepsilon_0}\left(\exp\frac{e\phi(x)}{k_B T} - \exp\frac{-e\phi(x)}{k_B T}\right) \tag{2.19}$$

For low potentials ($e\phi \ll k_{\mathrm{B}}T$) a Taylor expansion yields the linearized Poisson–Boltzmann equation:

$$\frac{\mathrm{d}^2\phi}{\mathrm{d}x^2} = \frac{en_{\mathrm{i},0}}{\varepsilon\varepsilon_0}\left(1 + \frac{e\phi(x)}{k_{\mathrm{B}}T} - 1 + \frac{e\phi(x)}{k_{\mathrm{B}}T} \pm \cdots\right) \approx \frac{2 \cdot n_{\mathrm{i},0} \cdot e^2}{\varepsilon\varepsilon_0 \cdot k_{\mathrm{B}}T} \cdot \phi \tag{2.20}$$

The general solution of this differential equation is given as:

$$\phi(x) = C_1 \exp\frac{-x}{\lambda_{\mathrm{D}}} + C_2 \exp\frac{x}{\lambda_{\mathrm{D}}} \tag{2.21}$$

with

$$\lambda_{\mathrm{D}} = \sqrt{\frac{\varepsilon\varepsilon_0 \cdot k_{\mathrm{B}}T}{2 \cdot n_{\mathrm{i},0} \cdot e^2}} \tag{2.22}$$

the Debye length and C_1 and C_2 constants defined by boundary conditions. The boundary conditions include that the potential at the surface is the surface potential $\phi_0 = \phi(x = 0)$, as well as that the potential disappears for large distances from the surface $\phi(x \to \infty) = 0$. The latter directly leads to $C_2 = 0$, from the first BC we get $C_1 = \phi_0$. The potential decreases exponentially following:

$$\phi = \phi_0 \cdot \exp\frac{-x}{\lambda_{\mathrm{D}}} \tag{2.23}$$

The reader familiar with the Debye length for electrons in plasmas, $\lambda_{\mathrm{D,e}} = \sqrt{\frac{\varepsilon_0 \cdot k_{\mathrm{B}}T}{n_0 \cdot e}}$ may have noticed the differences. Here, the screening of the surface charges is due to positive and negative ions in a monovalent salt, thus both ions are taken into account. If ions of higher valency are present, the Debye length is given by:

$$\lambda_{\mathrm{D}} = \sqrt{\frac{\varepsilon\varepsilon_0 k_{\mathrm{B}}T}{e^2\sum_i n_{i,0}Z_i^2}} \tag{2.24}$$

with Z_i the valency of the ion i. The decay length increases with increasing ion density. Close to a charged surface, local ion densities, and thus local pH, can be significantly different from the bulk liquid. The high local ion densities will also impact the permittivity of the liquid. Water, for example, has an $\varepsilon \approx 6$ for the first layer of molecules and $\varepsilon \approx 30$ for the second layer of molecules [3].

The low-potential assumption to linearize the Poisson–Boltzmann equation (equation (2.18)) works well for many practical cases in colloidal science; however, for electrochemistry or plasma–liquid-interactions, $k_{\mathrm{B}}T \ll e\phi$, and we cannot linearize the Poisson–Boltzmann equation. Then, it is convenient to generally solve the one-dimensional Poisson–Boltzmann equation and to define a dimensionless potential $\xi \equiv e\phi/k_{\mathrm{B}}T$. For a monovalent 1:1 salt, the Poisson–Boltzmann equation can then be written as:

$$\frac{\mathrm{d}^2\xi}{\mathrm{d}x^2} = \frac{e^2 n_{\mathrm{i},0}}{\varepsilon\varepsilon_0 \cdot k_{\mathrm{B}}T}(\exp(\xi) - \exp(-\xi)) \tag{2.25}$$

The relation between surface potential ϕ and surface charge σ is given by the Grahame equation:

$$\sigma = \sqrt{8n_0\varepsilon\varepsilon_0 k_{\mathrm{B}}T}\,\sinh\left(\frac{e\phi_0}{2k_{\mathrm{B}}T}\right) \tag{2.26}$$

This equation can be derived easily, using the one-dimensional Poisson–Boltzmann equation and the electroneutrality condition, demanding that the total charge in the electric double layer must be zero $\sigma = -\int_0^{\infty} \rho_{\mathrm{e}}\mathrm{d}x$.

The Grahame equation is useful to determine the capacitance of the diffusive layer. The differential capacitance between two regions of separated charges is defined as: $\mathrm{d}Q/\mathrm{d}U$, with Q the charge and U the voltage. We can thus calculate the capacitance of the electric double layer as:

$$C_{\mathrm{d}}^{\mathrm{A}} = \frac{\mathrm{d}\sigma}{\mathrm{d}\phi_0} = \sqrt{\frac{2e^2 n_0\varepsilon\varepsilon_0}{k_{\mathrm{B}}T}}\cosh\left(\frac{e\phi_0}{2k_{\mathrm{B}}T}\right) = \frac{\varepsilon\varepsilon_0}{\lambda_{\mathrm{D}}}\cdot\cosh\left(\frac{e\phi_0}{2k_{\mathrm{B}}T}\right) \tag{2.27}$$

Interestingly, for small surface potentials when the cosh can be expanded into a series and only the first term is considered, the capacitance becomes:

$$C_{\mathrm{d}}^{\mathrm{A}} = \frac{\varepsilon\varepsilon_0}{\lambda_{\mathrm{D}}} \tag{2.28}$$

The capacitance of the electric double layer is nothing else but the capacitance of a plate capacitor where the distance between the plates is given by the Debye length. The capacity of the electric double layer and thus the ability to store charges rises with increasing salt concentration because the Debye length decreases.

Chapter 2—summary

- Surfaces and interfaces have properties distinct from the bulk, since boundary molecules lack full coordination.
- Solid surfaces exhibit atomic rearrangements, oxidation, and altered electronic structures that strongly affect interactions with plasma.
- The concept of surface energy links atomic-scale bonding with macroscopic observables like surface tension and stress.
- Liquid surfaces are dynamic, with surface tension, Laplace pressure, and curvature-dependent vapor pressure governing stability and exchange.
- Adsorption of solutes or surfactants alters surface tension (Gibbs isotherm), enabling control over interfacial behavior.
- Charged liquid interfaces form electric double layers, whose potential and capacitance depend on ion density and are central to plasma–liquid electrochemistry.

Reading recommendations
- Introductory books on interface science are, for example, [10, 11].
- The physics and chemistry of interfaces is described in more depth in [3].

References

[1] The Editors of Encyclopaedie 1998 Encyclopaedia britannica, interface https://www.britannica.com/science/surface-chemistry-and-physics Accessed: June 2024

[2] The Editors of Encyclopaedie 1998 Encyclopaedia britannica, interface https://www.britannica.com/science/interface-physics Accessed: June 2024

[3] Butt H -J, Graf K and Kappl M 2023 *Physics and Chemistry of Interfaces* (Wiley)

[4] Lüth H 2001 *Solid Surfaces, Interfaces, and Thin Films* (Springer)

[5] Tang F, Ohto T, Sun S, Rouxel J R, Imoto S, Backus E H G, Mukamel S, Bonn M and Nagata Y 2020 Molecular structure and modeling of water-air and ice-air interfaces monitored by sum-frequency generation *Chem. Rev.* **120** 3633–67

[6] Costanza J, Arshadi M, Abriola L M and Pennell K D 2019 Accumulation of pfoa and pfos at the air-water interface *Environ. Sci. Technol. Lett.* **6** 487–91

[7] Bhusal P *et al* 2018 Comparing human peritoneal fluid and phosphate-buffered saline for drug delivery: do we need bio-relevant media? *Drug Deliv. Transl. Res.* **8** 708–18

[8] Neira-Carrillo A, Zárate I A, Nieto E, Butto-Miranda N, Lobos-González L, Campo-Smith M D, Palacio D A and Urbano B F 2022 Electrospun poly (acrylic acid-co-4-styrene sulfonate) as potential drug-eluting scaffolds for targeted chemotherapeutic delivery systems on gastric (AGS) and breast (MDA-Mb-231) cancer cell lines *Nanomaterials* **12** 3903

[9] Thomson W 1872 On the equilibrium of vapour at a curved surface of liquid *Proc. R. Soc. Edinburgh* **7** 63–78

[10] Adamson A W and Gast A P 1997 *Physical Chemistry of Surfaces* (Wiley)

[11] Evans D F and Wennerström H 1999 The Colloidal Domain: Where Physics, Chemistry *Biology, and Technology Meet* (Wiley)

IOP Publishing

Katharina Stapelmann

Chapter 3

Introduction to plasma surface interactions

3.1 Plasma at atmospheric pressure

At-a-glance
Non-equilibrium atmospheric pressure plasmas can roughly be divided into streamer and Townsend discharges. Streamers form when electron avalanches exceed a threshold, creating rapidly moving ionization fronts. Their propagation and stability depend on pulse timing, polarity, and gas composition. Townsend discharges, with sufficient ionization, will develop into a glow discharge. Direct plasmas expose surfaces to charged particles and short-lived species, while indirect plasmas deliver only long-lived species and photons. This distinction governs the type of chemistry and electric coupling at the interface.

3.1.1 Introduction

This eBook focuses on the interactions of plasmas with surfaces and it is assumed that the reader is familiar with the basic concepts of a plasma. For a concise introduction to low-temperature plasma physics, the reader may consult [1] and for atmospheric pressure non-equilibrium plasmas [2]. Atmospheric pressure plasmas in non-equilibrium, also called non-thermal or low-temperature plasmas, have been used since the 19th century, with one of the oldest experiments dating back to Siemens' ozone generation with a dielectric barrier discharge in 1857. Indeed, ozone formation went on to be one of the most successful early applications of atmospheric pressure plasmas together with electrostatic precipitation, combustion, and surface functionalization [2, 3]. In the early 2000s, atmospheric pressure plasma applications experienced a 'renaissance', with new applications emerging in medicine, agriculture, and water treatment. The ability to treat heat-sensitive material and liquids with these types of discharges paved the way for numerous applications where the plasma (or more general, the discharge, as not all of the non-equilibrium discharges actually qualify as

doi:10.1088/978-0-7503-6254-2ch3

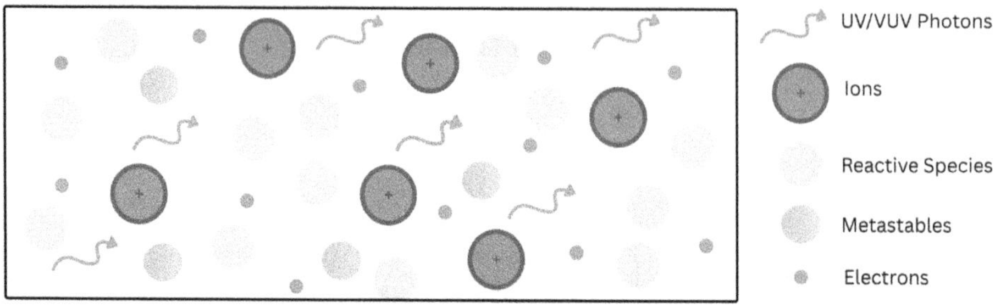

Figure 3.1. Plasma is a cocktail of active ingredients.

a plasma) is interfacing a solid or a liquid surface. The surfaces can be rather complex, as for example, living cells, animals, or even humans can be treated with the plasma. The surface can consist of multiple components and react to and interact with the interfacing plasma. The plasma itself is a rather complex cocktail of active ingredients, see figure 3.1, consisting of UV and VUV photons, ions, reactive species, metastables, and electrons. Depending on the discharge type, not all components will reach the treated object (discussed in more detail in section 3.1.3 below).

Non-equilibrium atmospheric pressure plasmas can roughly be divided into *streamer discharges* and *Townsend discharges*. As usual in plasmas, it all starts with an electron. The initial electron is accelerated to energies sufficient for ionization. New ion–electron pairs are created in collisions, resulting in an electron avalanche. In 1940, Meek formulated: 'A streamer will develop when the radial field about the positive space charge in an electron avalanche attains a value of the order of the external applied field' [4], also known as the Meek criterion. This is attained when the amplification of the avalanche reaches a critical number of electrons N_{cr} in the avalanche head which is around $N_{cr} \approx 10^8$. When the amplification of the avalanche reaches N_{cr} before arriving at the opposite electrode, local space charge accumulation leads to the *streamer breakdown*. A thin plasma channel propagates by ionizing the gas in front of its charged head due to the strong electric field induced by the streamer head itself. If the number of electrons in the avalanche head stays below the critical value N_{cr}, a Townsend discharge develops that, with sufficient ionization, will develop into a glow discharge. An essential feature of the Townsend criterion is that the space charge of a single electron avalanche does not distort the applied (homogeneous) electric field in the gap. Breakdown is reached when the current amplification in a gap tends to infinity, also known as the Townsend criterion for stationary self-sustainment, if the current amplification in a homogeneous field is:

$$I = I_0 \frac{\exp(\alpha d)}{1 - \gamma(\exp(\alpha d) - 1)} \tag{3.1}$$

with α the ionization coefficient and γ the secondary electron emission coefficient. At atmospheric pressure, Townsend or glow discharges tend to develop instabilities that allow the transition to a spark discharge. More information on this topic and the transitions occurring in atmospheric pressure plasmas can be found in [2] and references therein.

3.1.2 Streamer discharges

If the Meek criterion is fulfilled, a streamer develops. Streamer are transient discharges that only exist for a short period of time on the order of 10s of ns. If the streamer channel is able to connect the electrodes, electric breakdown occurs and a spark develops on the order of 100s of ns. If more power and in particular current is available, an arc will develop on the order of 500 ns to microseconds. Several methods can be used to avoid the streamer to spark and arc transition, e.g., to limit the power and current available to the discharge through the power supply or through electric circuit elements; to supply voltage pulses shorter than the time required to develop a spark (⩽100–300 ns), or to introduce a dielectric barrier on one or both electrodes.

In streamer discharges, the plasma channel propagates by ionizing the gas in front of its charged head (see figure 3.2 for an illustration of Dawson's photoionization theory [5]). (1) The positive streamer head consists of positive ions in its ionization front, emitting high-energetic photons. (2) The photons produce photo-electrons, and the high electric field between the positive ions and the electrons accelerates the electrons. (3) The accelerated electrons generate new ion–electron pairs in collisions with the background gas. (4) If enough electrons are produced by the photons, they neutralize the positive ions in the streamer head. (5) A new positive ionization front emits photons and the process starts all over again, (6) leading to the propagation of the streamer. The moving ionization fronts lead to the appearance of 'plasma bullets'. Short exposure time images of these discharges have shown high emission intensities in the moving ionization front, followed by a dimmer, almost dark, quasineutral plasma channel. More details can be found in, e.g., [6, 7].

Different discharge parameters affect the propagation of the ionization fronts. Shorter pulse rise times increase the plasma plume length in plasma jets [8]. For shorter pulse rise times, plasma current and breakdown voltage increase, as well as the electron temperature. This leads to a higher propagation velocity and a longer plume length. For pulse widths in the 100s of nanoseconds, increasing the pulse width also leads to an extended plasma plume, as more energy can be coupled into the plasma and the externally applied electric field is affecting the propagation. An

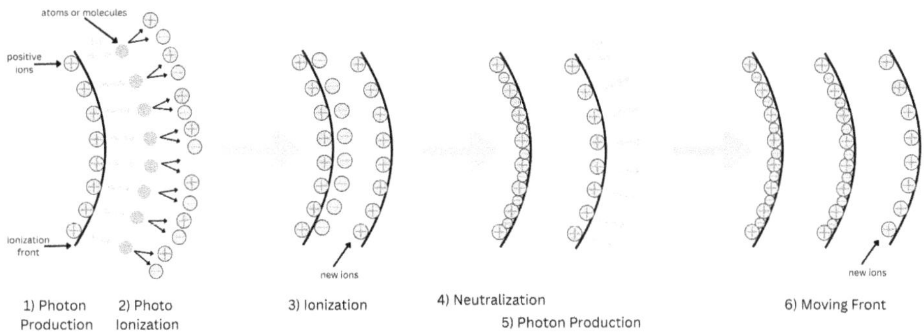

Figure 3.2. Streamer propagation based on photoionization and moving ionization fronts.

increase beyond 1000 ns does not effect the plasma plume [6]. The polarity of the voltage pulse has an impact on the plume length as well, as soon as the plume is exiting the nozzle. When the plasma bullet exits the nozzle, more air entrainment provides nitrogen that is easily ionized. In case of the positive applied voltage, the internal electric field is in the same direction of the externally applied electric field, enhancing the electric field. In case of the negative voltage pulse, the internal electric field is opposing the externally applied electric field, thus weakening the electric field. As a result, positive voltage pulses yield higher propagation velocities and longer plasma plumes than negative voltage pulses [7]. The plume length scales linearly with the amplitude of the applied voltage, as long as there is a uniform gas flow [9]. However, most jets are operated in noble gases entering surrounding air, so that streamer propagation will eventually be inhibited by higher air entrainment.

With all the parameters affecting the plume length, plasma jets can be created that either touch or do not touch the interfacing substrate [10], impacting the interactions of the plasma with the substrate. At this point it seems appropriate to distinguish between different treatment modalities. The plasma can be interfacing the surface with all active ingredients, which we define as direct plasma source here. If the surface is only exposed to a plasma effluent, where no charged particles are present and the short-lived species are decaying, we will refer to them as indirect plasma sources. This should not be confused with direct or indirect treatment, which is often used to distinguish between the direct plasma treatment of a substrate and the treatment of a liquid which is then used to treat the substrate.

3.1.3 Direct versus indirect plasma sources

The distinction between direct and indirect plasma sources becomes necessary when we are looking at the plasma components that are present at an interface. Direct plasma sources include, for example, volume dielectric barrier discharges (DBDs) and certain plasma jet types. The substrate that is treated becomes part of the electric circuit [11], see the illustration in figure 3.3, left. In this case, the substrate is exposed

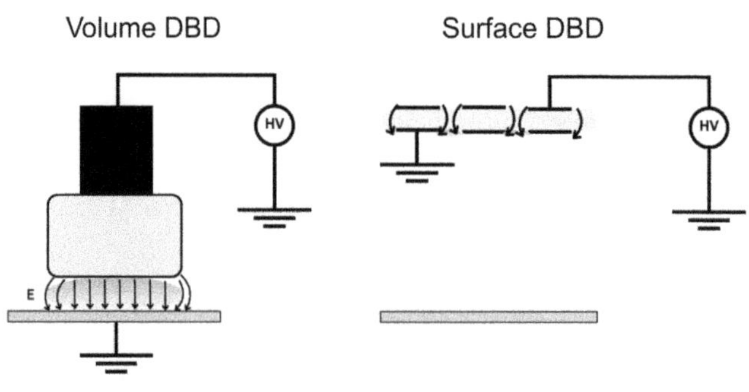

Figure 3.3. Illustration of volume (right) and surface (left) dielectric barrier discharges as examples for direct plasma sources and indirect plasma sources. The treated object becomes part of the electric circuit in case of the volume DBD and the active plasma is in direct contact with the treated object. In case of the surface DBD, the treated object is not part of the electric circuit and the active plasma is at some distance from the treated object.

to the high voltage, current flow, and charged species. Short-lived species are produced in the plasma in vicinity to the substrate. In indirect plasma sources, for example surface DBDs (see figure 3.3, right) or remote jet types, the substrate is not exposed to the high voltage, current flow, and charged species. Short-lived species decay on their way from the active plasma to the substrate and the chemistry is changing towards more long-lived species. High-energetic species like metastables and (V)UV photons may be quenched and absorbed on the way, depending on the process gas and the surrounding atmosphere as well as the distance between plasma and the substrate. With that in mind, we can now explore the components at interfaces and their impact on a surface.

3.2 Plasma components at interfaces

At-a-glance
Gas phase chemistry: Electron-driven excitation, dissociation, and ionization create short-lived species that can evolve into longer-lived molecules such as ozone, with product distribution depending on discharge type. Charged species and the plasma sheath: Electrons and ions form a sheath at surfaces, setting ion fluxes and energies. Even though ion energies are low at atmospheric pressure, transient spikes and sheath effects influence charging and reactions. Furthermore, the sheath provides the boundary conditions for the discharge and impacts bi-directional mass transport. UV/VUV photons and metastables: UV/VUV photons can break surface bonds, while metastables provide long-lived energy reservoirs that sustain ionization or activate surfaces through processes like Penning ionization.

3.2.1 Gas phase chemistry

Plasma chemistry is complex and its own subfield of study. This section shall by no means be exhaustive and only provide a general overview of gas-phase chemistry relevant for the purpose of this eBook. For a more detailed dive into plasma chemistry, this textbook [12] is a good resource, and for atmospheric pressure plasmas in air [13] is recommended for further reading.

The non-equilibrium conditions in low-temperature plasmas allow for energetic chemical processes to occur at (or close to) room temperature. As always in plasmas, it starts with an electron. The high-energetic electrons collide with atoms or molecules and depending on their energy, different processes occur:

- excitation: $X + e^- \rightarrow X^* + e^-$
- dissociation: $X_2 + e^- \rightarrow X + X + e^-$
- ionization: $X + e^- \rightarrow X^+ + e^- + e^-$
- attachment: $X + e^- \rightarrow X^-$

During electron impact collisions species can get excited and if the energy is sufficient, ionization occurs. In molecular gases the excitation can occur on an electronic level like in atoms; in addition, molecules can undergo vibrational and rotational excitation. Another feature of molecular gases is the possibility for dissociation, which may occur due to electron impact directly, or by climbing a vibrational ladder where low-energetic electrons can excite the molecule vibrationally in a stepwise manner which eventually leads to the dissociation of the molecule. Once the molecule is dissociated, the dissociation partners form new bonds with other reaction partners. One example is the formation of ozone in air or oxygen-containing plasmas:

$$O_2 + e^- \rightarrow O + O + e^- \tag{3.2}$$

$$O + O_2 + M \rightarrow O_3 + M \tag{3.3}$$

At first (equation (3.2)), molecular oxygen is dissociated by electron impact collision. The resulting atomic oxygen can react with molecular oxygen to form ozone (equation (3.3)). For this reaction, a third body (denoted as 'M') is necessary for momentum and energy conservation. This is one example why we have to distinguish between direct and indirect plasma sources. In a direct plasma source, atomic oxygen is created in vicinity to the surface and can interact with the surface. The further away the plasma source is from the surface, the less atomic oxygen and the more ozone (or other longer-lived species) will be present at the interface.

For chemical reactions, ground-state neutrals, ions, and metastable states (see section 3.2.3) are important. Short-lived excited states carry energy but they also decay radiatively via optically allowed dipole transitions on a timescale of nanoseconds ($t_{\text{rad}} \approx$ ns). The short residence time makes their contributions to the chemistry negligible (unless their lifetime is comparable to the inverse collision frequency) [13]. Electron-induced dissociative ionization and dissociation processes take nanoseconds to microseconds ($t_{\text{diss}} \approx$ ns–µs). Reactions between neutral species at atmospheric pressure take milliseconds to seconds ($t_{\text{reaction}} \approx$ ms–s) and when free radicals are involved microseconds to milliseconds ($t_{\text{radicals}} \approx$ µs–ms). The reactions can be described with a rate coefficient k. For two-body electron collision processes, this rate coefficient depends on the energy of the electron. Each process has a threshold energy E_{th} below which the process is energetically not possible. This threshold as well as the energy-dependent probability for a electron collision process to occur is contained in the cross-section $\sigma(v)$. Electrons in a plasma are not mono-energetic but follow an energy distribution function $f(E)$ or velocity distribution function $f(v)$. The rate coefficient can then be expressed as

$$k(v) = \int \sigma(v) v f(v)\, dv \tag{3.4}$$

The velocity v in equation (3.4) refers to the relative velocity between the colliding particles. As the electrons have significantly higher velocities than the heavy particles, which can be treated at rest relative to the fast moving electrons, the velocity is essentially the electron velocity. For collisions between heavy particles, which are considered randomly moving, the reaction probability is determined by the relative velocity. The velocity distribution (at equilibrium) is dependent on the heavy particle temperature, T. The temperature-dependent rate coefficient takes the form of a modified Arrhenius equation:

$$k(T) = AT^n \exp \frac{E_a}{k_B} T \tag{3.5}$$

where A is a the Arrhenius factor, E_a is the chemical reaction activation energy, k_B is the Boltzmann constant, and n is a curvature parameter ($-1 \leqslant n \leqslant 1$).

All gas-phase reactions start with an electron, as mentioned above. Electron-driven reactions create excited states, ions, and can lead to dissociation, following the rate coefficient described in equation (3.4). These reactions are then followed by heavy particle processes following the rate coefficient described in equation (3.5). In the active plasma where electrons are present, both types of reactions occur. In the effluent or afterglow of a discharge, i.e., in pulsed plasmas during the off-phase or for indirect plasma sources like surface DBDs or certain plasma jets, the gas phase chemistry close to the target is dominated by heavy particle reactions.

Once the species are produced, we are interested in the flux reaching the substrate. A fundamental transport process for neutrals is diffusion. Starting with the momentum equation for the species of interest to explore the velocity first, including the collisional term on the right given by the momentum transfered during a collision ($m\vec{v}$) times the frequency of occurence ($n\nu_m$):

$$mn\left(\frac{\partial}{\partial t}\vec{v} + (\vec{v}\ \nabla\ \vec{v})\right) - qn\vec{E} - \nabla\ p - mn\vec{v}\,\nu_m \tag{3.6}$$

where m is the mass of the species of interest, n their density, $\vec{v}$ the velocity, q elementary charge, $\vec{E}$ the electric field, ∇p the pressure gradient, and ν_m the momentum transfer collision frequency. Under steady-state conditions the temporal derivative can be set to zero. For the case of highly collisional atmospheric pressure plasmas, the second term on the left can be neglected and equation (3.6) simplifies to:

$$\vec{v} = \frac{1}{mn\nu_m} \cdot (qn\vec{E} - \nabla\ p) \tag{3.7}$$

Considering the ideal gas law, two cases can be identified. For isothermal processes:

$$\nabla p = k_B T\ \nabla\ n \tag{3.8}$$

and for adiabatic processes:

$$\frac{\nabla p}{p} = \gamma \frac{\nabla n}{n} \tag{3.9}$$

where γ is dependent on the degrees of freedom of the system. For plasma processes, we usually assume an isothermal process, so that equation (3.7) can be written as:

$$\vec{v} = \frac{q}{m\nu_m}\vec{E} - \frac{k_B T}{m\nu_m}\frac{\nabla n}{n} \tag{3.10}$$

Multiplying the velocity with the density yields the flux of species:

$$\Gamma = n\vec{v} = n\frac{q}{m\nu_{\mathrm{m}}}\vec{E} - \frac{k_{\mathrm{B}}T}{m\nu_{\mathrm{m}}}\nabla n \tag{3.11}$$

The first term on the right-hand side describes the drift caused by the electric field (for charged species) and the second term describes the diffusion due to a density gradient. Considering now an interface, solid or liquid, we are interested in the flux of species reaching the interface. To determine the flux of species, we need the mean velocity of the particles in the gas volume, which we obtain by averaging the velocity distribution function:

$$\langle v \rangle = \int v f \, \mathrm{d}^3 v \tag{3.12}$$

Assuming a Maxwellian distribution function:

$$f(\vec{v}) = \left(\frac{m}{2\pi k_{\mathrm{B}}T}\right)^{\frac{3}{2}} \cdot \exp - \frac{mv^2}{2k_{\mathrm{B}}T} \tag{3.13}$$

The mean velocity results in:

$$\langle v \rangle = \left(\frac{8k_{\mathrm{B}}T}{\pi m}\right)^{\frac{1}{2}} \tag{3.14}$$

with the mean velocity in one direction:

$$|\langle v_x \rangle| = \left(\frac{2k_{\mathrm{B}}T}{\pi m}\right)^{\frac{1}{2}} = \frac{1}{2}\langle v_x \rangle \tag{3.15}$$

The flux of species in the direction of the substrate, here in positive x direction, is then:

$$\Gamma = n\frac{1}{2} \cdot \frac{1}{2}\langle v \rangle = \frac{1}{4}n\langle v \rangle \tag{3.16}$$

The flux changes when the substrate becomes a sink of species. Considering unreactive species and a solid substrate, the particles are just reflected and the boundary condition is $n = 0$ at the substrate. There is no variation in density. Considering reactive species, or a liquid interface, a particle loss due to reactions or transport through the interface needs to be considered. This is done by defining the reflection coefficient r, with $0 < r < 1$. Let us imagine an interface within the distance of the mean free path λ from the surface. The flux is then given by a component directed towards the wall Γ_{out} and a component directed away from the wall Γ_{in}. The difference between these fluxes equals the part of the flux that is lost to the walls, Γ_{loss}:

$$\Gamma_{\mathrm{loss}} = \Gamma_{\mathrm{out}} - \Gamma_{\mathrm{in}} \tag{3.17}$$

The density of the species at the position of the imagined interface can be calculated based on the flux equations, using equation (3.15) for the velocity:

$$n = \frac{1}{\frac{1}{2}\langle v \rangle}(\Gamma_{\text{out}} + \Gamma_{\text{in}}) \tag{3.18}$$

The flux Γ_{out} is correlated to the flux Γ_{in} via the reflection coefficient r:

$$\Gamma_{\text{in}} = r\Gamma_{\text{out}} \tag{3.19}$$

This closes the set of equations, as we have three equations (3.17)–(3.19) for three unknowns. Solving for Γ_{loss} yields:

$$\Gamma_{\text{loss}} = \frac{1}{2}nv \cdot \frac{1-r}{1+r} = \frac{1}{2}nv \cdot \frac{\beta}{2-\beta} \tag{3.20}$$

where β is the surface loss probability with $\beta = 1 - r$.

3.2.2 Charged species and the plasma sheath

The life-sustaining process in a plasma is inelastic collisions of electrons and atoms or molecules that lead to ionization, as introduced in section 3.2.1:

$$X + e^- \rightarrow X^+ + e^- + e^- \tag{3.21}$$

where X could be any atom or molecule. The collision creates an electron–ion pair. The electrons are much lighter than the ions ($m_e = 9.109 \times 10^{-31}$ kg and $m_{i,H} = 1.673 \times 10^{-27}$ kg for the lightest ion H^+) and thus, also more mobile than the ions. Free electrons can reach a surface by diffusion on a much shorter timescale than ions will. This leads to negative charging of the surface (if it is on floating potential) and an electric field is created, repelling other negative charges and attracting positive charges towards the surface. This concept is commonly known as the *plasma sheath* and was first introduced by Langmuir in 1923 [14]. The acceleration of ions towards the surface can have undesirable effects, e.g., damage to the surface or an increased heat flux, or it can be used for the application, e.g., plasma etching and sputtering. For further reading on plasma sheath, see, e.g., [15–17].

Less research has been done on plasma sheaths in atmospheric pressure plasmas. The collisionality of atmospheric pressure plasmas and the sheath will prevent ions from getting accelerated to significant energies like in low-pressure plasmas, and from hitting the surface with the anisotropy that can be achieved in low-pressure conditions where the ion mean free path is larger than the sheath width. Ions in the sheath will undergo many inelastic collisions before they reach the substrate, where the mean ion energies are typically below 1 eV [18]. Computational work by Babaeva and Kushner has shown the importance of the substrate properties [19–21]. Ion energies observed ranged from below 1 eV with a few eV in the high-energetic tail of the ion energy distribution function (IEDF) for small particles to 10–15 eV for thick dielectric layers up to 200 eV for streamer penetration into dielectric pores. The high energies were transient, lasting for a few nanoseconds. This transient energization of the ions is attributed to the dynamic compression of the plasma sheath between a streamer head and the surface, as the streamer is approaching it. The sheath and the surface layer act as capacitive voltage divider. The resulting electric field depends strongly on the dielectric properties of the surface, where high-permittivity material is preserving the sheath voltage longer due to their higher capacitance, allowing more efficient ion acceleration. Low-permittivity material on the other hand experiences faster voltage collapse, limiting ion energies. Recently, the first experimental investigation of the ion energy distribution function from the impact of an atmospheric dielectric barrier discharge plasma jet on surfaces was performed [22]. In this study, time-resolved molecular beam mass spectrometry was used to monitor the IEDF. It was found that all ions reach the spectrometer at less than 1 eV. However, the time-resolution of the spectrometer would not have allowed for a detection of higher-energetic ions at nanosecond timescales. To conclude, the ion energies reaching a substrate at atmospheric pressure typically have low energies, with potentially some very high energies for a very brief period of time. Nonetheless, the sheath sets the boundary condition for the plasma and will thus influence the discharge. The sheath contributes to the interfacial physics and chemistry, and to bi-directional mass transport if the surface is a liquid [23], as we will see in section 5.3. It will also provide high-energetic particles to the surface (cf section 3.3.2), even if they are less high-energetic and directed compared to their low-pressure plasma siblings. Charged particles as well as the sheath only play a role in direct plasma sources interacting with surfaces and will be discussed more in sections 4.2, 4.4, and 5.3.

3.2.3 UV/VUV photons and metastables

Atoms and molecules gain energy during inelastic collisions with electrons which can lead to the population of electronically excited states, as mentioned above. An example for an energy level diagram is shown in figure 3.4 for the helium atom.

Excited states have a very short lifetime on the order of nanoseconds. Excessive energy is released during a transition from an upper to a lower level in form of spontaneous emission of a photon with:

$$E = h \cdot f \tag{3.22}$$

where E is the energy, h is Planck's constant, and f is the frequency, or wavelength ($\lambda = c/f$). The energy is equivalent to the ΔE of the upper and lower energy level and is distinct for each atom or molecule, thus providing a fingerprint of the plasma. Furthermore, many transitions occur within ΔE of 3–12 eV corresponding to photons in the UV and VUV range, see table 3.1. The shorter the wavelength, the higher the energy but the lower the penetration depth. Radiation in the UV and in the VUV can break virtually any existing bond. The role of UV and VUV radiation can be very different for different plasma discharges. Discharges in air, for example, do not show high levels of UV radiation and most of the radiation is within the UV-A. VUV radiation is absorbed rapidly by oxygen, so that surrounding air or

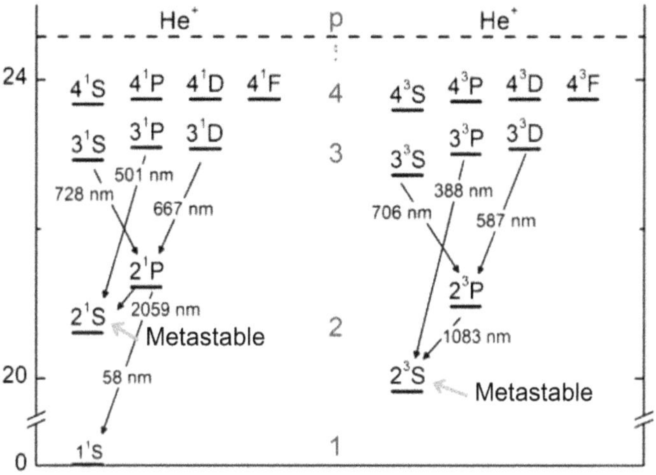

Figure 3.4. Helium energy levels and corresponding wavelength for de-excitation from excited states as an example. Adapted from [24]. CC BY 4.0.

Table 3.1. Ultraviolet radiation with wavelength and energy ranges.

	Wavelength	Energy	Name
UV-A	380–315 nm	3.26–3.94 eV	Soft UV
UV-B	315–280 nm	3.94–4.43 eV	Intermediate UV
UV-C	280–200 nm	4.43–6.20 eV	Hard UV
VUV	100–200 nm	6.20–12.4 eV	Vacuum UV

discharges in air will not provide large amounts of VUV radiation. If plasma jets are used that are operated with a gas flow of helium or argon, the carrier gas can provide an optical channel in which UV and VUV photons can travel un-absorbed to the substrate. Plasma jets in helium or argon can generate significant fluxes of VUV radiation, which can induce oxidation of surface carbon groups [18], or initiate reactions on PFAS via photolysis [25], for example. However, measurements of VUV at atmospheric pressure is very difficult and not much reported in literature. Some VUV measurements in controlled atmosphere have been performed on the COST jet in helium and argon gas mixtures [26].

Not all transitions between energy levels are allowed; some transitions are forbidden based on quantum mechanical selection rules. Energy levels from which no emitting transition is allowed are called *metastable states*. The helium atom has two metastable states, see figure 3.4. Metastable states have very long lifetimes (low Einstein coefficients), up to minutes, and can play an important role in plasma discharges. The He(2^3S) and He(2^1S) metastable states have threshold energies of 4–5 eV below the ionization energy (see figure 3.4). This leads to the preferential formation of metastables over the He^+ ion in low-temperature discharges. The metastables can be necessary to sustain a plasma. Stepwise ionization, or Penning ionization, can be the dominant ionization mechanism in a plasma. This is often the case for helium discharges, as the ionization energy of helium is high. The role of metastables in surface interactions is less explored and deserves more attention in future research. We will discuss metastable and UV/VUV interactions with solids and liquids in sections 4.3 and 5.4, respectively.

3.3 Surface kinetics

At-a-glance
Neutrals can adsorb, scatter, or react at surfaces. Adsorption strength varies from weak physisorption to strong chemisorption, influencing subsequent reactions. Plasma surface chemistry follows either Eley–Rideal (direct impact) or Langmuir–Hinshelwood (adsorbed–adsorbed) mechanisms, with both contributing to interfacial reactivity. Although less energetic than in vacuum plasmas, ions, photons, and metastables still drive bond cleavage, photoemission, and surface activation at atmospheric pressure.

In the last section, the plasma components at interfaces were introduced. The components can roughly be divided into *low-energy* and *high-energy* particles, discussed in more detail in the following subsections.

3.3.1 Low-energy particles

The typical gas temperatures of low-temperature atmospheric pressure plasmas are in the range of $T_g \approx 300$–2000 K, corresponding to a kinetic energy of $E_k \approx 0.026$–0.17 eV. This is lower than the typical binding energy between atoms in the material the plasma is interfacing. Any low-energy particle, typically a neutral atom or molecule, approaching the interface can undergo a reaction or be reflected, depending on the particle's reactivity. Figure 3.5 outlines the possible interactions [27]. The incoming particle can be reflected with the probability r, react at the surface to recombine to a volatile product with the probability γ, or it can stick to the surface with the sticking probability s. The overall surface reaction probability is defined as: $\beta = s + \gamma = 1 - r$.

In the absence of a gas flow, the flux Γ of gas molecules approaching the surface is proportional to the gas pressure and given by:

$$\Gamma\left(\frac{\text{molecules}}{\text{cm}^2\ \text{s}}\right) = \frac{N_A p}{\sqrt{2\pi m R T}} = 3.52 \times 10^{22}\ \frac{p\ [\text{in Torr}]}{\sqrt{mT}} \tag{3.23}$$

with N_A the Avogrado's number, m molecular weight of the gas, p the pressure, R the gas constant, and T the temperature. At atmospheric pressure and room temperature, the fluxes of different molecular weights are in the range of 10^{23} molecules cm^{-2} s^{-1}. It takes about 10^{15} molecules cm^{-2} to cover a surface with a

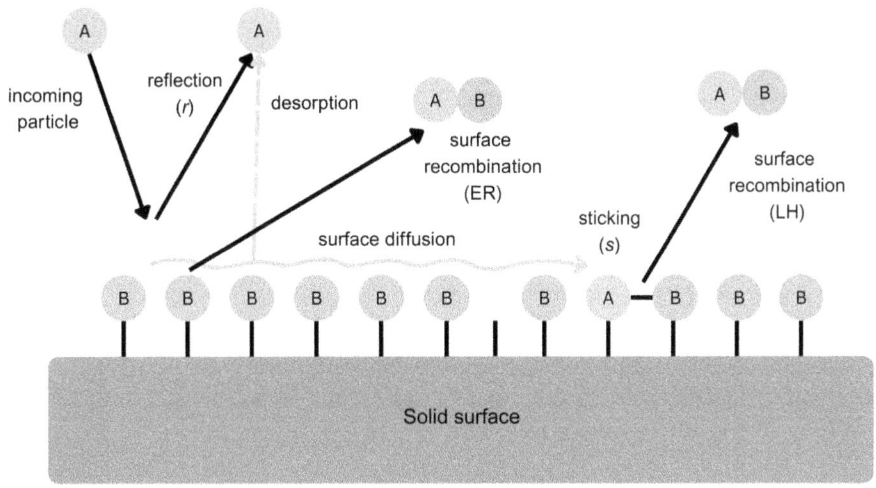

Figure 3.5. Low-energy particles interacting with a surface. Figure recreated based on [27], copyright IOP Publishing Ltd. All rights reserved.

unimolecular layer [28], see also section 2.2. Hence, a surface exposed to a gas, will be covered immediately if every incident molecule would stick to the surface. Adsorption is a spontaneous process; it occurs because it reduces the imbalance of forces acting on surface atoms and molecules of a solid or a liquid. The energetics are essentially the same for all interfaces; the distribution of excess energy, however, will be heterogeneous for solid surfaces while for liquids it is assumed that adsorption is a uniform process [29]. The amount adsorbed is determined by different factors, among them the partial pressure for solids and the concentration for liquids. Once adsorbed, the rotational and vibrational freedom of adsorbed molecules is reduced and even their electrical properties can change. Molecules can diffuse laterally or react chemically, or might desorb again into the gaseous phase.

A gas atom or molecule experiences an attractive potential when approaching the surface and the strength of this potential determines the interaction between the gas and the surface atoms. The forces involved in adsorption processes are nonspecific van der Waals forces, ionic or electrostatic forces, and specific forces involved in the formation of chemical bonds [29]. Specific forces are orders of magnitude larger than the nonspecific forces. Interactions with specific forces are termed *chemisorption*, whereas the weaker nonspecific forces based interactions are termed *physisorption*. In the absence of complicating factors (e.g., capillary condensation), physisorption is diffusion controlled and occurs as rapidly as adsorptives[1] can be delivered to the surface. The sublimation energy is around 20–40 kJ mol^{-1} ($\approx$0.2–0.4 eV) and the adsorbate[2] can still diffuse relatively freely on the surface. Most of the physisorption occurs due to weak van der Waals dipole–dipole interaction where the potential

[1] Adsorptive: the substance to be adsorbed, before it is on the surface.
[2] Adsorbate: the material in the adsorbed state.

energy scales with $E_{\text{pot}} \approx \frac{1}{r^6}$. The molecular structure of the solid does not change in general (with a few exceptions). Equilibrium establishes quickly. In the case of chemisorption, the adsorbtion energy is on the order of the chemical binding energies ($\leqslant 0.5$ eV). A chemical bond is formed in conjunction with an electron exchange that can be donated from the conduction band in metals. For insulators, the reaction will occur at an unpaired electron, or at a radical site/dangling bond, or a radical can be inserted into an existing bond. The adsorbate is relatively immobile and is not free to diffuse on the surface. If the molecules are able to penetrate the surface layer and diffuse into the bulk, the term *absorption* is used.

When is a particle approaching the surface now captured and when reflected? For that, we need to investigate the energy. The particle loses a part p_{loss} of its kinetic energy E in the collision with the surface, dependent on its mass m and the mass of the surface atoms M. It is captured when:

$$E_{\text{after}} = (1 - p_{\text{loss}}) \cdot E_{\text{before}} \leqslant E_{\text{a}} \tag{3.24}$$

where E_{a} is the depth of the potential wall the particle sees when approaching the surface. Maximum energy transfer is reached when $m = M$, following $4 \cdot \frac{m \cdot M}{(m+M)^2}$. This reduces to $4 \cdot m/M$ when $m \ll M$. The sticking coefficient is higher when

- the kinetic energy E is small;
- the depth of the potential wall E_{a} is large;
- the masses m and M are similar and the energy transfer during collision is high.

3.3.1.1 Surface reactions

Once the particle is captured, surface reactions can occur. There are two simple theorems to describe surface reactions, indicated in figure 3.5. The first, Eley–Rideal (ER), describes the direct reaction upon impact. This reaction scheme only works for barrier-less and exothermic reactions and it usually involves radicals. An example would be hydrogen abstraction from a surface: H(g) + H(s) $\rightarrow$ H_2(g), where *g* denotes gaseous phase and *s* denotes solid phase. The reaction rate is given by:

$$R = k_{\text{AB}} \cdot p_{\text{A}} \cdot \Theta_{\text{B}} \tag{3.25}$$

where k_{AB} is the rate constant for the reaction, p_{A} gives the partial pressure of component A, related to the flux of particles to the surface, and Θ_{B} denotes the surface coverage of component B on the surface. The second process is the Langmuir–Hinshelwood (LH) model which is the most common surface reaction mechanism. The reaction occurs among adsorbates on the surface. Here, particle A is first adsorbed, diffuses along the surface, and finds a partner B to react with. The model allows reactions between molecules as well. Here, the reaction rate is also depending on the surface coverage but this time of both reactants:

$$R = k_{\text{AB}} \cdot \Theta_{\text{A}} \cdot \Theta_{\text{B}} \tag{3.26}$$

An example would be the reaction of CO(s) + O(s) $\rightarrow$ CO_2.

3.3.2 High-energy particles

At atmospheric pressure, plasma species that are considered high-energy, such as ions, UV/VUV photons, and metastable species, interact with surfaces in distinctive ways due to the high collision frequency and limited mean free path. A summary of high-energy particles, their energy range when approaching a surface, and their main effects on the surface is given in table 3.2. Ions, as described in section 3.2.2, reach the surface with kinetic energies of around 1 eV. Higher energies are possible depending on the substrate material, but those higher energies are transient on the order of nanoseconds. Nonetheless, even at reduced energies, ions can drive secondary electron emission which influences sheath dynamics and plasma properties. Local electric fields can push ions towards the surface, enhancing surface charging. For liquid substrates, this can potentially modify electrical double layers. Despite their low mean energy, the ions deliver energy locally and can participate in ion-assisted reactions or help desorb species from the surface, e.g., weakly bound intermediates or contaminants. The energy flux can result in localized heating and in case of a liquid surface to enhanced evaporation, also called ion-enhanced evaporation. Ion impact on the surface leads to momentum transfer, influencing adsorbates. This can be more pronounced in conjunction with surface radicals or excited species. Ions may not play a major role in atmospheric pressure plasmas interacting with surfaces but can be seen as helpers or enablers for certain surface processes.

UV and VUV photons have the advantage of being collisionless. If not absorbed on the way, they will reach the surface with their full energy. VUV has very limited penetration depth in air, so the effects of VUV are local to the near-surface region unless the photons can travel in an optically transparent channel. Once the photons reach the surface, they can directly break molecular bonds in surface contaminants or adsorbates, leading to photodissociation, photooxidation, or photolysis. When combined with reactive species, the photons can cleave a bond and a radical reacts with the fragment. Especially VUV photons can induce photoemission of electrons, charging surfaces or enabling surface reactions with transient electrons.

Metastables can carry substantial energy (10–20 eV) and, with their long lifetimes of μs–ms, can reach the surface if not quenched on the way. At the surface, metastables transfer their energy to the surface. This can lead to an energy transfer to adsorbates, causing desorption or activation. The provided energy can help to overcome reaction barriers. If quenched to the ground state at the surface, the

Table 3.2. High-energy particles and their main surface effects.

Component	Energy at surface	Main surface effects
Ions	$\leqslant$1 eV	Assist reactions, charge surfaces, secondary electron emission
UV/VUV	3–12 eV	Bond breaking, surface activation, photoemission
Metastables	10–20 eV	Energy transfer, Penning ionization

transferred energy can also be released as heat or secondary electrons. Lastly, Penning ionization can occur where metastables can ionize surface-adsorbed molecules or impurities. Metastables are a non-ionizing but potent source of chemical and electronic surface activation.

Chapter 3—summary

- Atmospheric pressure plasmas are non-equilibrium systems producing electrons, ions, photons, and radicals that interact strongly with surfaces.
- Streamer dynamics, gas composition, and applied pulse characteristics control stability and plume propagation.
- A key distinction exists between direct plasma sources (exposing surfaces to charged and short-lived species) and indirect sources (delivering only long-lived species and photons).
- Gas-phase chemistry is initiated by electron-driven collisions and evolves toward longer-lived molecules such as ozone.
- Plasma sheaths define ion fluxes and energies, with transient spikes influencing charging and interfacial chemistry.
- UV/VUV photons and metastables provide additional activation channels through bond cleavage, ionization, and Penning processes.
- Surface interactions are governed by adsorption and reaction mechanisms, ranging from physisorption to chemisorption, and follow Eley–Rideal or Langmuir–Hinshelwood pathways.

Reading recommendations

- For an introduction to atmospheric pressure plasmas, e.g., 'Foundations of low-temperature plasma physics' [1] and 'Foundations of atmospheric pressure plasmas' [2]. For a deeper dive, the textbook *Non-Equilibrium Air Plasmas at Atmospheric Pressure* is recommended [13].
- Streamer discharge physics in [30].
- For a deeper dive into plasma chemistry, this textbook [12] is a good resource, and for atmospheric pressure plasmas in air [13] is recommended for further reading. A detailed analysis of long timescale chemistry in air plasmas interfacing liquid-covered tissue is presented in [31].
- Photoionization, ionization fronts, and plasma bullets: [6, 7].
- Plasma sheath, general theory: [15–17] and for plasma–liquid interactions: [23, 32].
- Surface functionalization is nicely introduced in 'Foundations of plasma surface functionalization of polymers for industrial and biological applications' [18].

References

[1] Keudell A V and Der Gathen V S-V 2017 Foundations of low-temperature plasma physics–an introduction *Plasma Sources Sci. Technol.* **26** 113001

[2] Bruggeman P J, Iza F and Brandenburg R 2017 Foundations of atmospheric pressure non-equilibrium plasmas *Plasma Sources Sci. Technol.* **26** 123002

[3] Brandenburg R, Becker K H and Weltmann K D 2023 Barrier discharges in science and technology since 2003: a tribute and update *Plasma Chem. Plasma Process.* **43** 1303–34

[4] Meek J M 1940 A theory of spark discharge *Phys. Rev.* **57** 722

[5] Dawson G A and Winn W P 1965 A model for streamer propagation *Z. Geophys.* **183** 159–71

[6] Lu X, Reuter S, Laroussi M and Liu D 2019 *Nonequilibrium Atmospheric Pressure Plasma Jets: Fundamentals, Diagnostics, and Medical Applications* (CRC Press)

[7] Lu X, Laroussi M and Puech V 2012 On atmospheric-pressure non-equilibrium plasma jets and plasma bullets *Plasma Sources Sci. Technol.* **21** 034005

[8] Wu S, Xu H, Lu X and Pan Y 2013 Effect of pulse rising time of pulse DC voltage on atmospheric pressure non-equilibrium plasma *Plasmas Process. Polym.* **10** 136–40

[9] Walsh J L, Olszewski P and Bradley J W 2012 The manipulation of atmospheric pressure dielectric barrier plasma jets *Plasma Sources Sci. Technol.* **21** 034007

[10] Norberg S A, Tian W, Johnsen E and Kushner M J 2014 Atmospheric pressure plasma jets interacting with liquid covered tissue: touching and not-touching the liquid *J. Phys. D: Appl. Phys.* **47** 475203

[11] Stapelmann K, Gershman S and Miller V 2024 Plasma-liquid interactions in the presence of organic matter–a perspective *J. Appl. Phys.* **135** 160901

[12] Fridman A 2008 *Plasma Chemistry* (Cambridge University Press)

[13] Becker K H, Kogelschatz U, Schoenbach K H and Barker R J 2004 *Non-Equilibrium Air Plasmas at Atmospheric Pressure* (CRC Press)

[14] Langmuir I 1923 Positive ion currents from the positive column of mercury arcs *Science* **58** 290–1

[15] Robertson S 2013 Sheaths in laboratory and space plasmas *Plasma Phys. Control. Fusion* **55** 093001

[16] Riemann K U 2008 Plasma and sheath *Plasma Sources Sci. Technol.* **18** 014006

[17] Franklin R N 2003 The plasma-sheath boundary region *J. Phys. D: Appl. Phys.* **36** R309

[18] Booth J P, Mozetič M, Nikiforov A and Oehr C 2022 Foundations of plasma surface functionalization of polymers for industrial and biological applications *Plasma Sources Sci. Technol.* **31** 103001

[19] Babaeva N Y and Kushner M J 2011 Ion energy and angular distributions onto polymer surfaces delivered by dielectric barrier discharge filaments in air: I. Flat surfaces *Plasma Sources Sci. Technol.* **20** 035017

[20] Babaeva N Y and Kushner M J 2011 Ion energy and angular distributions onto polymer surfaces delivered by dielectric barrier discharge filaments in air: II. Particles *Plasma Sources Sci. Technol.* **20** 035018

[21] Babaeva N Y and Kushner M J 2013 Control of ion activation energy delivered to tissue and sensitive materials in atmospheric pressure plasmas using thin porous dielectric sheets *J. Phys. D: Appl. Phys.* **46** 125201

[22] Henze D, Chauvet L and von Keudell A 2025 Ion energy distributions from the impact of an atmospheric dielectric barrier discharge plasma jet on surfaces arXiv:2505.07771

[23] Vanraes P and Bogaerts A 2021 The essential role of the plasma sheath in plasma-liquid interaction and its applications–a perspective *J. Appl. Phys.* **129** 220901

[24] Wünderlich D and Fantz U 2016 Evaluation of State-Resolved Reaction Probabilities and Their Application in Population Models for He, H, and H2 *Atoms* **4** 26

[25] Jeon J, Dias T C and Kushner M J 2025 Plasma degradation of surfactant molecules at the gas–liquid interface *ISPC Conf. Abstract* https://www.ispc-conference.org/files/boa/paper/238.pdf

[26] Golda J, Biskup B, Layes V, Winzer T and Benedikt J 2020 Vacuum ultraviolet spectroscopy of cold atmospheric pressure plasma jets *Plasma Process Polym.* **17** 1900216

[27] Benedikt J 2010 Plasma-chemical reactions: low pressure acetylene plasmas *J. Phys. D: Appl. Phys.* **43** 043001

[28] Somorjai G A 1972 *Principles of Surface Chemistry* (Prentice-Hall)

[29] Myers D 1991 *Surfaces, Interfaces, and Colloids* (Wiley)

[30] Nijdam S, Teunissen J and Ebert U 2020 The physics of streamer discharge phenomena *Plasma Sources Sci. Technol.* **29** 103001

[31] Lietz A M and Kushner M J 2016 Air plasma treatment of liquid covered tissue: long timescale chemistry *J. Phys. D: Appl. Phys.* **49** 425204

[32] Meyer M, Nayak G, Bruggeman P J and Kushner M J 2022 Sheath formation around a dielectric droplet in a He atmospheric pressure plasma *J. Appl. Phys.* **132** 083303

IOP Publishing

Atmospheric Pressure Plasmas at Interfaces

Katharina Stapelmann

Chapter 4

Plasma–solid interactions

Plasma–solid interactions are central to plasma-based material processing, surface modification, and a wide range of emerging applications. These interactions are governed by a complex interplay of chemical, physical, and electrical processes occurring at the plasma–surface interface. Depending on the plasma conditions and the nature of the target material, different mechanisms, such as reactive species delivery, ion bombardment, photon-induced reactions, and energy transfer, may dominate. Table 4.1 summarizes the main plasma–solid interaction mechanisms, the key species involved, and typical surface effects, serving as a reference for the detailed discussions that follow in this chapter.

Table 4.1. Summary of plasma–solid interaction mechanisms, plasma species, and typical surface effects.

Category	Plasma species	Typical surface effects
Neutral species (Section 4.1)	Reactive radicals, neutral atoms/ molecules	Material removal, surface functionalization, oxidation, reduction, deposition
Charged species (Section 4.2)	Ions, electrons	Surface charging, defect creation, secondary electron emission
Photons (Section 4.3)	UV, VUV	Photochemical reactions, bond scission, desorption
Metastables (Section 4.3)	Metastable atoms/ molecules	Energy transfer, reaction sensitization, surface excitation, secondary electron emission
Plasma-induced flow (Section 4.4)	EHD flow, ionic wind	Enhanced transport of reactive species, boundary layer modification, heat transfer
Thermal effects (section 4.4)	Localized heating by plasma	Increased surface temperature, enhanced reaction kinetics, possible thermal damage

doi:10.1088/978-0-7503-6254-2ch4

4.1 Neutral species at the interface

At-a-glance
Reactive neutrals can clean, etch, or functionalize surfaces, and also enable thin-film growth via processes such as PECVD and ALD at atmospheric pressure.

Atmospheric pressure plasmas are an efficient source of reactive species that can interact with a solid interfacing the plasma or the plasma effluent. The fate of the neutral approaching the surface depends on many factors, most importantly the available energy. Excited species, radicals, or neutrals with a high impact energy will be more reactive than thermal species which are often just reflected. The surface itself plays a role as well; high temperature surfaces are more reactive than cold surfaces. The combination of reactive species produced in the plasma and the material the solid is made of determines the reactivity between the incoming particle and the solid substrate. For example, if the solid is a hydrogenated amorphous carbon layer and the plasma sustained in a helium/oxygen gas mixture, in which atomic oxygen is one of the dominating reactive species, the atomic oxygen will react with the carbon and produce volatile CO or CO_2, thus etching the material [1]. On the other hand, if an air plasma is interfacing an aluminum substrate, the reactivity will be low as the aluminum (covered by a thin oxide layer) will not react with the reactive species produced in the plasma. In the following sections, material removal, surface functionalization, and deposition will be introduced as examples for neutral interactions with solid surfaces.

4.1.1 Material removal

Among the reactive species produced in a plasma, the hydroxyl radical OH and atomic oxygen O are probably the most reactive and unselective oxidative species. Etching, the removal of bulk material, is one of the potential outcomes. In the example mentioned above, an etching rate of about 1×10^{15} C atoms s^{-1} was observed [1]. Surfaces can also be cleaned with plasma. Cleaning refers to the removal of material located on the surface, for example, the removal of hydrocarbon containing lubricants and other fatty contaminants [2]. The underlying chemical processes are very similar for both, etching and cleaning. As described in section 2.2, the surface of a solid material is different from the bulk material. Depending on the material and its environmental exposure, the physical and chemical composition of a surface can be complex and vastly different from the bulk material. Non-uniform layers of adsorbed water and organic or inorganic contamination can be found on different surfaces. The surface roughness can add to the complexity. Surface cleaning removes naturally formed or process-introduced contamination on surfaces, often to prepare a surface for surface treatments such as protective or functional coatings, more discussed in the next section. Boselli *et al* used a plasma jet and a dielectric barrier discharge (DBD) device to demonstrate the feasibility of using plasma to effectively clean the surface of a corroded 19th century daguerreotype [3]. An important parameter for surface cleaning is the specific surface energy which can be determined by contact angle measurements. After plasma cleaning, surface tension is typically higher (in particular for DBDs). This effect is temporary (unless the surface is functionalized, see below); a high wettability remains stable for about 24 h if the energy dosage of the DBD is between 50 and 100 J cm^{-2} [2].

4.1.2 Surface functionalization

The functionalization of surfaces has been widely used to improve adhesion properties of materials, e.g., for better printing, coating, and bonding. Non-stick surfaces can be glued and bonded following plasma treatment [4]. Plastic foils, fibers, and other polymer materials often have non-polar surfaces that are chemically inert with surface energies in the 20–40 mN m^{-1} range. Polymers are wetted by a liquid when the surface energy of the polymer exceeds the surface energy of the liquid. Common organic solvents have lower surface energies, e.g., toluene ($\sim$28 mN m^{-1}) or ethanol ($\sim$22 mN m^{-1}), so that paint and inks based on organic solvents can be applied to plastics. Water-based inks, preferred due to environmental and safety concerns, require pre-treatment of the surface as water has a high surface energy ($\sim$72 mN m^{-1}). The low surface energy of plastics helps to impede surface contamination and allows for easy cleaning, but it complicates printing as the sticking probability is low. Using plasma, the surface layer can be functionalized to increase the surface energy. This is often achieved by inserting polar groups at the surface. While different polymers and different gas discharges may produce different results, the example of polypropylene (PP) functionalization by a humid air plasma illustrates the general procedure which is to a certain extent applicable to a variety of different materials and gas discharge types. Reactive species produced by the plasma show different reactivities towards PP. In the model developed by Dorai and Kushner [5], it was found that O, H, OH, and O_3 are the most important reactive species. The reactivity of the reactive species with PP differs depending on the position of the C atom, with the highest found to be for C atoms that are only bound with one C atom and the lowest for those that are bound with three C atoms inside the polymer. The surface reactions can be classified in three categories: (i) initiation, (ii) propagation, and (iii) termination. Initially, the PP surface is free of any oxygen. As it is true for many organic materials, the initiation reaction is the H abstraction by O or OH, the subscript *g* denotes gaseous:

(i) initiation:

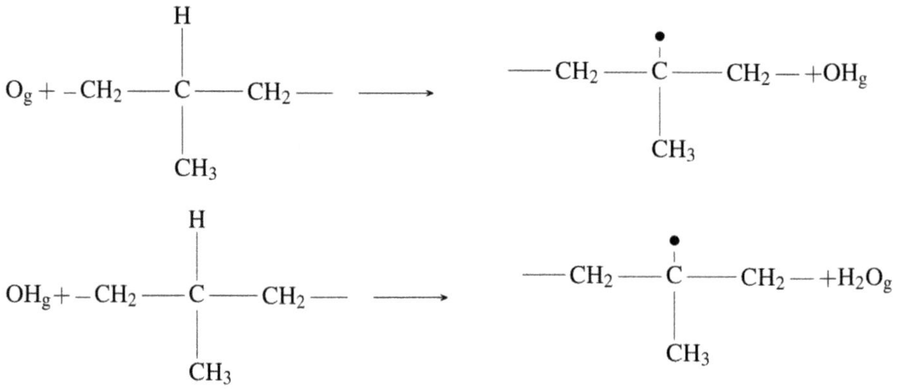

In both cases, an alkyl radical is generated.

(ii) propagation:

The alkyl radical on the PP surface can react with O_2 to form peroxy radicals, or with O or O_3 to form alkoxy radicals:

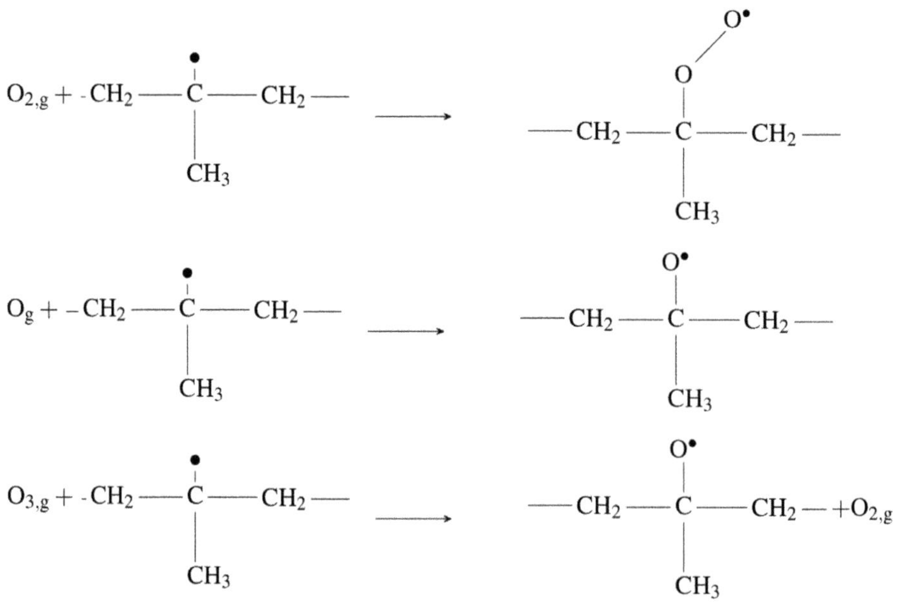

The created peroxy radicals on the surface can react further by abstracting neighboring H atoms, creating more alkyl radicals:

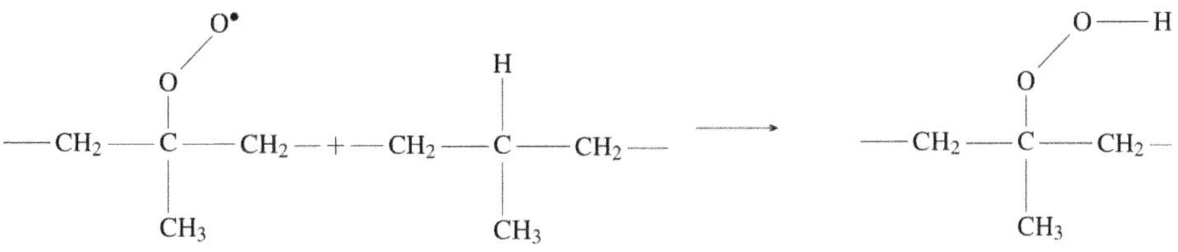

During oxidation, chain backbone scission occurs with the alkoxy radical being the main source, leading to the formation of ketones or aldehydes:

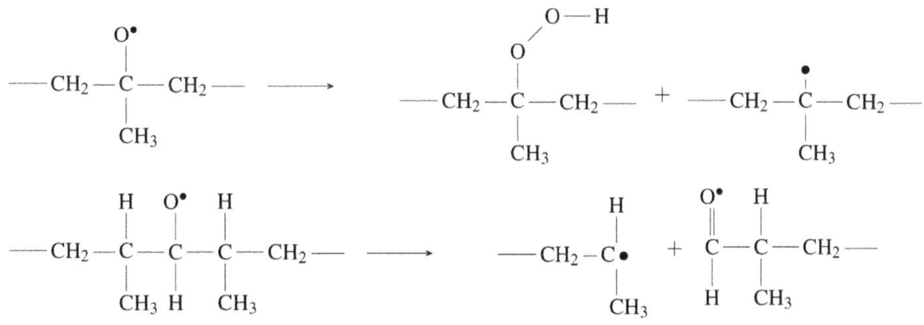

Alkoxy radicals reacting with PP can also form alcohol groups:

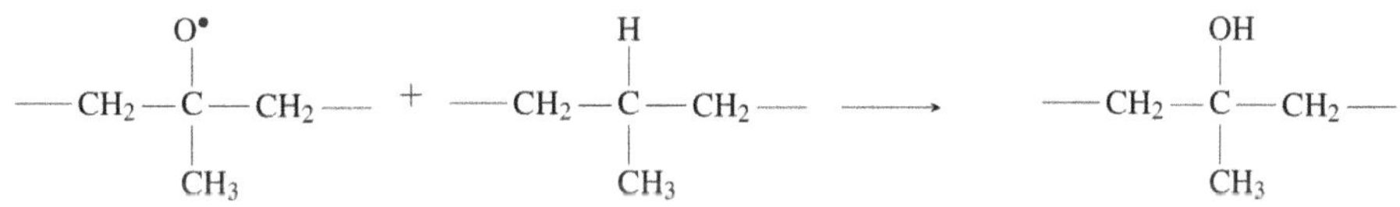

The alcohol groups can react further with OH and O and form alkoxy radicals:

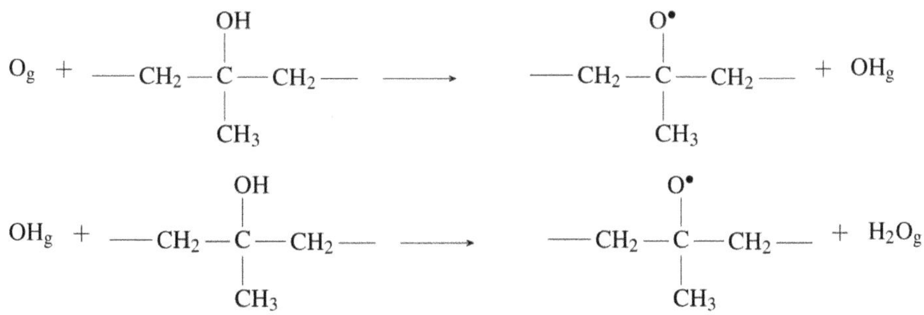

(iii) termination:

Reactions of H and OH with alkyl and carbonyl radicals result in the formation of alcohols and acids that are more stable than the free radicals and are thus seen as stable end products, terminating the reaction:

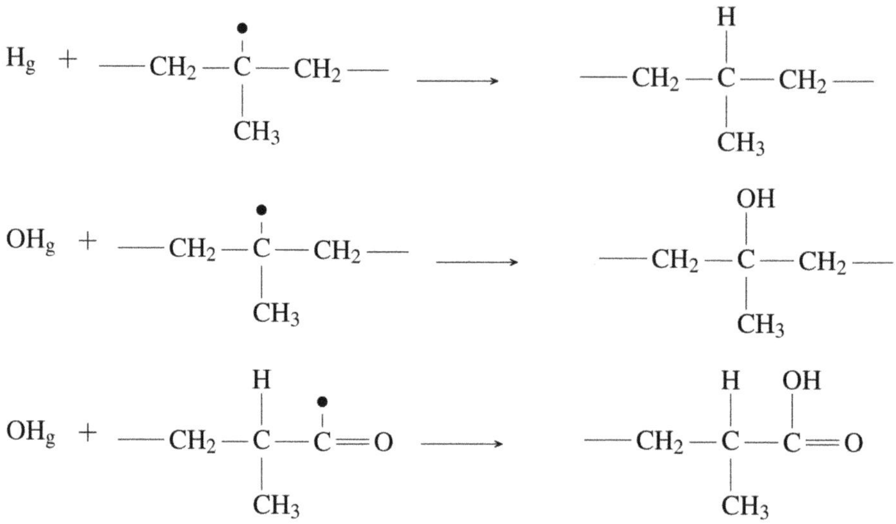

These reactions illustrate some of the typical pathways to be expected when organic molecules are exposed to plasma. While different materials and different discharge types exhibit specific pathways, the general procedure of initiation, propagation, and termination holds true independently. The initiation by H abstraction, followed by propagation of intermediate and instable products, and termination in more stable products, has also been observed for the amino acid cysteine [6] and the protein glutathione [7], as examples. Further reading: Booth *et al* 'Foundations of plasma surface functionalization of polymers for industrial and biological applications' [8].

4.1.3 Deposition

Plasma-enhanced chemical vapor deposition (PECVD) has been used for decades in industry to achieve thin film depositions of various materials for multiple applications, typically under low-pressure conditions. With the rise of atmospheric pressure plasmas, deposition without the need for expensive vacuum equipment becomes an interesting alternative. Plasma provides reactive species that promote surface reactions and high-energetic electrons to dissociate a variety of feedstock gases and gas/liquid precursors. Deposited films can include oxides, nitrides, carbides, and organosilicon compounds. Plasma parameters such as power and gas composition can be tuned to influence film properties like density, adhesion, and stress. PECVD under atmospheric pressure conditions works very similar to the low-pressure PECVD process. A precursor is transferred by a carrier gas, typically inert, and mixed with the feed gas, before entering the discharge. In the discharge, high-energetic electrons dissociate the precursor.

Reactive species are formed from feed gas and precursor and directed to the substrate where deposition occurs. With the use of atmospheric pressure conditions, in-line roll-to-roll web processing systems for flexible materials and sheet-type processing systems for rigid materials can be used for continuous thin film deposition.

Organic vapor precursors, such as hexamethyldisiloxan (HMDSO) or tetraethyorthosilicate (TEOS), can be used to deposit SiO_2 films under atmospheric conditions. Examples include anti-reflective coatings, in food packaging as moisture or oxygen barrier, and dielectric coatings [9]. Titanium dioxide (TiO_2) is another example for growth of oxide films under atmospheric pressure conditions. TiO_2 coatings have a wide range of applications, from electronics to solar cells and photocatalysis [10]. The most widely used precursor for TiO_2 coatings is titanium tetra-isopropoxide (TTIP). Other precursors include, for example, titanium ethoxide (TEOT) and titanium bis(acetylacetonate) diisoprpoxide (TIPO) [10]. Deposition rates of up to 40 $\mu m\ s^{-1}$ were reported when using aerosolized TTIP with an open air atmospheric pressure plasma jet [11]. Further examples are provided, e.g., in [12], where PECVD of fluoropolymers by a DBD is described for thin film deposition on outer and inner surfaces of complex 3D porous material, the deposition of multicomponent hydrocarbon polymer/ZnO nanoparticles nanocomposite thin films, as well as a DBD jet deposition of thin films containing carboxylic acid groups. Another review is provided in [13], focusing on devices for large area coatings, microplasmas for localized coatings, and exploring the correlation between power dissipation and deposition rate and quality of the film.

Atomic layer deposition (ALD) is another example for a deposition process well-known from low-pressure plasmas, that has been adapted to be used under atmospheric pressure conditions. ALD is a self-limiting, sequential deposition technique that allows for atomic-scale control over film thickness and composition. The precursor only reacts with available surface functional groups, leading to precise, layer-by-layer growth. Under low-pressure conditions this is typically achieved by alternating exposures of the substrate to different precursors, separated by purge or evacuation steps. Under atmospheric pressure conditions, spatial ALD is often used where precursors are delivered in different zones.

A microplasma printer combining chemical vapor deposition and ALD was developed and employed for the deposition of thin film patterns of TiO_2 [14]. With this setup, a one-step patterning approach for thin oxide films is possible by depositing on selected areas of a substrate according to a software-defined pattern. The microplasma printer is a multi-needle-to-plate dielectric barrier discharge and was originally developed for surface treatment and functionalization of surfaces [15, 16]. Area-selective ALD of In_2O_3:H using the microplasma printer is described in [17] and the deposition of amine-containing organo-silane polymers using 3-aminopropyl trimethoxysilane as precursor in [18].

4.2 Charged species at the interface

At-a-glance
Ions and electrons interact with surfaces by charging, secondary electron emission, or inducing discharges; their fluxes are typically low but can locally alter plasma–surface dynamics.

As was described in section 3.1.3 and we will discuss more in section 4.4.1, charged species will only reach the interface when a direct plasma source is used. Thus, this section is focusing on direct plasma sources with a focus on streamer discharges.

The transient streamer discharges consist of propagating ionization fronts with self-organized field enhancements at their streamer heads [19], cf section 3.1.2. For many applications, streamer discharges are interfacing solid or liquid interfaces (for liquid interfaces cf section 5.3), e.g., in plasma medicine or plasma agriculture. The interaction of streamer discharges with dielectric and conducting surfaces involves complex mechanisms such as electrostatic attraction, secondary electron emission due to ion or metastable impact or high-energetic photons, and surface charging [19]. When a streamer is approaching a surface, the current reaches a local maximum due to a reconfiguration of the electric field. Electron production and loss mechanisms are balanced, leading to slowing down and eventually stopping the streamer propagation [20]. The volume streamer is deflected and the propagation continues along the dielectric surface. Ion-induced secondary electron emission can cause a secondary breakdown in the gap between streamer head and dielectric.

As briefly introduced in section 3.2.2, a computational study has shown that in heads of filaments, or streamer heads, the electric field can exceed 200 kV cm^{-1} when the head is far from a surface [21]. The electric field is then compressed into the voltage drop across the head when it is approaching a dielectric surface, as the conductivity of the streamer tail is a high-conductive plasma channel. In the resulting sheath, the electric field can exceed 400–800 kV cm^{-1}, depending on the permittivity of the dielectric surface (larger permittivities correspond to larger electric fields) [21]. Although the mean free path of ions in atmospheric pressure conditions is small (0.05–1 μm) [21, 22], the ions can accelerate in the high electric field along their short mean free path to energies ranging from 20 eV for low permittivity material and up to 100 eV for dielectrics with high permittivity [21]. As the intersection of streamer with the dielectric is transient, the acceleration of ions during that phase is limited to a few nanoseconds. Nonetheless, with activation energies for chemical reactions on the order of a few eV and sputtering thresholds of some tens of eV, the ions could provide sufficient energy. However, the flux of high-energetic ions is small and limited in time.

In another computational study, streamers propagating towards dielectric surfaces as well as conductive surfaces with and without secondary electron emission was studied [23]. The surface properties significantly influence the streamer dynamics, particularly the axial propagation velocity and the sheath thickness. Secondary electron emission enhances streamer propagation and alters sheath characteristics. In their study, electric fields of 177 kV cm^{-1} (sheath thickness s = 0.2 mm) were observed for a conductive surface with no secondary electron emission, and 550 kV cm^{-1} (s = 0.01 mm) for conductive surfaces with a secondary electron emission coefficient of $\gamma = 0.1$. For dielectric surfaces it was observed that the secondary electron emission is noticeable at low surface charge densities but diminishes when the charge density increases. As the permittivity of the dielectric was increased, the flux of ions to the surface increased [23].

The above cases describe positive streamers interacting with a surface. In negative streamers, the streamer head as well as the sheath region that is formed near the surface have lower electric field values. Fast secondary electrons are produced and trapped in the narrow anodic sheath, leading to a thin layer of precursor electrons ahead of the streamer [24]. In negative ionization waves, the energy of ions is close to thermal [25].

When jets are interfacing dielectric material (e.g., $\varepsilon_r \approx 2$ for most plastics, $\varepsilon_r \approx 5$ for tissue), surface charges are deposited on the dielectric. The electric field radially propagates inside the dielectric with lower electric field strengths than in axial direction [26] and low values of permittivity encourage propagation of the electric field into the dielectric, as well as the formation of surface ionization waves propagating along the surface of the dielectric [27]. Mean electron energies in the range of 3–8 eV were found in helium/air mixtures that remained constant during the radial surface wave propagation [28]. When the plasma is interfacing the dielectric, the electric field within the dielectric is influenced by two sequential effects. Initially, a relatively strong electric field arises from the vicinity of the ionization front and the high volume charge density ρ. Over longer timescales, lower electric fields are induced as a result of both residual volume charges near the dielectric and surface charges that accumulate on its surface [26]. Dielectrics with high permittivity promote a return stroke of the ionization wave, enabling the formation of a conductive channel between the plasma and the dielectric surface [27]. Furthermore, the higher capacitance resulting from the higher ε_r facilitates charge accumulation at the surface, promoting higher electric fields at the surface of the dielectric [27]. Experimental investigations of surface charges, in particular transient accumulation and decay of surface charges, are rather challenging. Recent advances in diagnostic methods and results are extensively reviewed in [29].

When a jet is interfacing a grounded metal (essentially infinite dielectric constant), the speed of the ionization wave, the magnitude of the ionization source, and the peak electric field are increasing, compared to material with lower permittivity such as dielectrics [27]. Once the ionization wave impacts the surface, no surface ionization wave occurs as there is no horizontal component of the electric field at the surface of the metal, and, of course, also no charging of the surface in case of a grounded metal plate. The applied voltage remains between the gap which in turn leads to a fully sustained conduction channel and a return stroke of the ionization wave that eventually reconnects with the electrode [27]. The direct path to ground for the conduction current and the lack of charge accumulation on the surface prevents electric field enhancement at the surface. However, there is a (much lower) electric field produced at the metal surface, originating in an anode sheath created by the accumulation of positive ions [27].

Lastly, ions approaching a solid interface can neutralize surface charges or cause secondary electron emission by Auger emission.

4.3 Photon and metastable-induced interactions

At-a-glance
UV/VUV photons and metastables transfer energy by breaking bonds, ionizing, or releasing electrons; these processes influence both surface chemistry and discharge propagation.

Metastables and photons both provide high energies. When metastables and photons are introduced onto a solid surface, the energy transfer to the surface can lead to local heating, direct breaking of bonds, and activation of surface sites (see also section 3.3.2). Metastable atoms have an extremely high probability of de-excitation ($\approx$1) upon contact with solid surfaces [30]. Their internal energy is released through electronic interactions. For metals and semiconductors, resonance ionization and Auger neutralization are the main processes. During resonance ionization, the outer electron of the metastable atom tunnels into an empty energy level of the solid. An electron from the solid fills then the inner orbital of the ionized metastable atom, while another electron is ejected from the solid (Auger neutralization). If no suitable empty state is available, such as in insulators, the metastable is quenched by Auger de-excitation: a surface electron fills the inner orbital of the metastable atom, and its outer electron is ejected. Alternatively, Penning ionization can occur. Similarly, when high-energetic photons are introduced onto a solid surface, photoemission releases electrons. These electrons can influence the discharge, and the propagation of volume and surface streamers. However, for all these processes, secondary electron emission coefficients are poorly known [31]. The influence of photoemission for photoemission coefficients of $\gamma_{\text{photon}} = 5 \times 10^{-4}$, $\gamma_{\text{photon}} = 5 \times 10^{-3}$, and $\gamma_{\text{photon}} = 1 \times 10^{-1}$ was studied computationally for a dielectric tube jet configuration in [31]. Including photoemission in the simulation increased the number of secondary electrons close to the dielectric surface and promoted the propagation of the discharge on the surface. As the photoemission coefficients increased, the secondary electron emission increased as well, leading to more electrons close to the dielectric surface. The discharge then propagated more easily on the surface and as a result, the discharge was observed to be more tubular [31]. In another computational study, comparisons were made including VUV and disregarding VUV photons. Vast differences in particle densities, fluxes, and even discharge current pulse amplitude were observed [32]. More research and combined experimental and computational approaches are needed to fully understand the interaction of UV/VUV photons and metastables with solid surfaces at atmospheric pressure.

4.4 Differences between direct and indirect plasma sources

At-a-glance
The regime-defining difference is whether the target participates in the electrical circuit, which sets sheath formation, charge transfer, and the partition of power at the interface. In *direct* sources, a sheath forms at the surface and ion/electron fluxes, together with transient charging/polarization, create strong feedback with the external circuit; substrate conductivity and permittivity therefore strongly tune discharge behavior. In *indirect* sources, interactions are mediated by neutrals, photons, and plasma-induced flow/thermal input, so nozzle–surface distance, ambient gas, and flow conditions govern species delivery, uniformity, and thermal dose.

Direct and indirect plasma sources differ in whether the target participates in the electrical circuit (see section 3.1.3), which in turn governs sheath formation, charge transfer, and how species, momentum, and heat reach the surface. Table 4.2 provides an overview of these differences and serves as a reference for the discussion

Table 4.2. Comparison of direct and indirect plasma discharges at a solid surface.

Aspect	Direct discharge	Indirect discharge
Electrical contact	Plasma is electrically coupled to the target	Plasma is not electrically coupled to the target
Sheath Formation	Forms at the target surface	Forms near plasma source or dielectric barrier
Charge transfer	Direct (current can flow to the surface)	Indirect (via displacement current or capacitive coupling), if any
Reactive species delivery	High-flux, often ion-driven	Diffusion- and convection-dominated
Energy transfer	Includes ion bombardment and electron flux	Limited to neutral species, photons, and thermal transfer
Surface charging	Controlled by target conductivity and plasma parameters	Influenced by capacitive effects and displacement currents, if any
Photon effects	Present, often secondary to charged species	Significant role in photochemical surface reactions
Plasma-induced flow	Typically minor unless intentionally induced (e.g., jets)	Often dominant in surface interaction (e.g., plasma jets)
Applications	Etching, deposition, direct surface activation	Gentle surface treatment, biomedical applications, remote activation

that follows. We first examine direct sources in section 4.4.1, where electrical coupling to the target sets up sheath-controlled dynamics and feedback between surface charging and discharge behavior (cf figure 4.1); we then turn to the indirect case, where interactions are mediated primarily by neutrals, photons, and plasma-induced flow with distance-dependent transport and thermal dosing effects (cf figures 4.2 and 4.3).

4.4.1 Direct plasma sources

Direct plasma sources, e.g., volume dielectric barrier discharges or certain plasma jet types, couple to the target electrically, providing a path for feedback and interactions. This section is based on a recent perspective paper by the author [33]. The electrical characteristics of a solid, such as its conductivity, dielectric constant, surface roughness, and surface energy, define the boundary conditions that govern electric field distributions near the surface. These boundary conditions control sheath formation, which subsequently influences plasma ignition, discharge dynamics, and the flux of charged and reactive species to the interface. The incoming fluxes modify the surface's chemical and physical state, while the evolving surface conditions, in turn, affect plasma behavior, establishing a dynamic feedback loop.

When a solid is exposed directly to plasma, it interacts with a variety of plasma-generated species, including reactive species, ions, electrons, and UV photons, cf chapter 3. These interactions result in surface charging, photoionization, secondary electron emission, quenching and de-excitation processes, and surface chemical reactions. The target's conductive properties, permittivity, and electrical grounding (or floating potential) play a key role in shaping these interactions. For example, metallic substrates can enhance radical formation in plasma jets [34]. The substrate also influences ionization front dynamics, secondary discharge formation, and surface-bound discharge propagation [35]. A comprehensive overview of plasma jets and surface interactions is provided in [36], where it is also noted that a lot of the underlying physics regarding sheath formation and development, as well as charge exchanges between plasmas and surfaces of different types are still unknown. When plasma interacts with dielectric or ferroelectric materials, the high electric fields in the sheath region can induce transient surface polarization. This polarization arises from field-driven alignment of bound charges, injection of charge carriers into surface states, or dipole reorientation in polarizable materials. In ferroelectrics, the sheath's electric field may temporarily switch or distort polarization domains, while in piezoelectric materials, plasma-induced mechanical or thermal stress can alter surface charge distribution. These effects are inherently dynamic and closely tied to surface charging and charge relaxation and can significantly influence local electric fields, sheath structure, and plasma–surface interactions. Transient polarization thus adds another layer of complexity to plasma–solid coupling, especially in applications involving dielectric barrier discharges, plasma treatment of polymers, or plasma-assisted functionalization of piezoelectric or ferroelectric surfaces.

Because electrons move much faster than ions, dielectric surfaces typically acquire a negative charge. Positive ions neutralize at the surface, sometimes causing secondary electron emission, while neutralized species may re-enter the gas phase. The net surface charge depends strongly on the material's dielectric constant. Modeling studies suggest that for materials with high dielectric constants (>4), electric fields within the material arise mainly from surface charges, while in low-dielectric materials, internal space charges can significantly contribute [37]. The dielectric constant also impacts the electric field distribution from a circuit perspective. A higher dielectric constant increases surface capacitance and reduces impedance, shifting more voltage drop, and thus electric field, into the plasma gap for a given applied voltage pulse. This affects discharge onset, propagation, and plasma parameters. Moreover, higher dielectric constants promote greater surface charge accumulation, intensifying return strokes in AC discharges during the opposing half-cycle and in pulsed discharges as the voltage falls [36, 37].

While microscopic models have advanced in simulating plasma interactions within the sheath and near-surface regions, they fall short in capturing system-level dynamics and changes of the entire system. Electrical circuit models may help bridge this gap, offering insights into discharge reproducibility, stability, and the influence of surface properties on plasma behavior [38]. An electrical circuit model approach should thus consider the entire system, including the substrate and pathways to ground. Figure 4.1 shows a selection of typical volume DBD and DBD jet setups and their respective equivalent circuits. The circuit of a volume DBD with both electrodes covered with a dielectric consists of the capacitors $C_{d1,2}$ for each dielectric

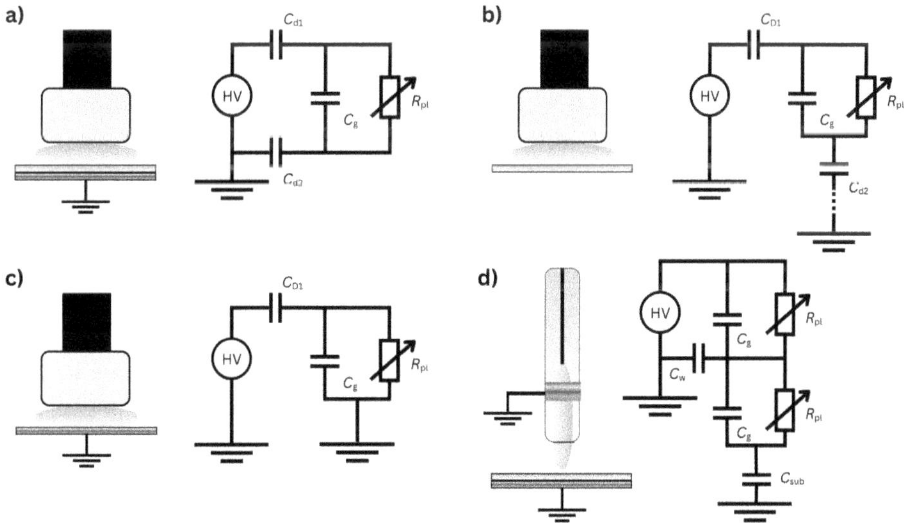

Figure 4.1. Selection of typical volume DBD and DBD jet setups and their respective equivalent circuits. The gas gap provides a capacitance C_g which is bridged by the variable resistance R_{pl} when the plasma ignites. (a) Double DBD: both electrodes are covered with a dielectric with the capacitance $C_{d,1}$ and $C_{d,2}$. (b) Floating DBD: the counter electrode is not electrically connected to a ground and is on floating potential. (c) The grounded electrode is directly exposed to the discharge, only the driven electrode is covered with a dielectric. (d) Configuration for a DBD jet: the dielectric wall material provides capacitance C_w, the substrate becomes part of the circuit as C_{sub}.

and the capacitance of the gas gap C_g, which is bridged by the variable resistance R_{pl} when the plasma ignites [39]. This approach clearly highlights the importance of the treated substrate, as the circuit changes when the grounded electrode is covered with a dielectric (figure 4.1(a)), and floating (figure 4.1(b)), or if the grounded metal electrode is directly exposed to the plasma (figure 4.1(c)). A similar approach can be used for plasma jets. Several different plasma jet configurations are described in the literature. Depicted in figure 4.1(d), the plasma is generated inside a dielectric tube with the capacitance of the wall material C_w and propagates to the treated substrate which has the capacitance C_{sub}. Regardless of the setup, changes in dielectric properties or conductivity of the treated solid will affect the circuit impedance and with that the power deposited in the plasma, as well as the power delivered to the substrate, setting up an electrical feedback loop. Electrical characterization of a plasma source should thus be performed as real-time electrical monitoring of the system with the substrate for each specific application.

4.4.2 Indirect plasma sources

Indirect plasma sources, e.g., surface DBDs or certain jet types like the COST jet [40], are not electrically coupled to the target. Nonetheless, the target can still interact with the plasma, the gas flow, or the plasma-induced gas flow. Plasma-induced gas flow has been extensively studied for flow control in plasma actuators, where typically surface DBDs are used to manipulate airflow or to introduce well-defined perturbations. Comprehensive reviews of these discharges for flow control applications is given, e.g., in [41, 42]. The interaction between charged particles and the electric field generates body forces within the gas volume. These forces give rise to electrohydrodynamic (EHD) flow, sometimes referred to as ionic wind, electric wind, or plasma-induced flow. EHD effects are driven by momentum transfer from drifting ions to neutral gas molecules. In regions where there is a net space charge, typically near the sheath or within ionization zones, ions accelerate under the influence of the electric field and impart momentum to the background gas via collisions. This can induce high-velocity gas streams, even in the absence of an external gas flow. The magnitude and pattern of plasma-induced flow depend on factors such as applied voltage, plasma geometry, carrier gas composition, and environmental conditions (e.g., pressure, humidity). While EHD forces are most prominent in atmospheric discharges with significant ion densities, even low-intensity flows can influence surface interactions by enhancing reactive species transport and boundary layer dynamics. Ionic winds have been extensively investigated for flow control, but have been neglected and underutilized in other applications, e.g., plasma medicine or plasma agriculture. Understanding plasma-induced flow is crucial when designing plasma sources for surface modification, as it governs not only how reactive species reach the substrate but also the uniformity and efficacy of the treatment process.

In plasma jets, gas flow and plasma dynamics can be tightly coupled. The forced gas flow carries plasma-generated reactive species toward the target surface, with EHD effects potentially further modifying local flow patterns near the surface.

At the stagnation point, where the jet impinges on the substrate, plasma-induced flow can thin the boundary layer, enhance mass and heat transfer, and facilitate species delivery to the surface. Flow recirculation zones and shear effects at the target may also impact local treatment uniformity and surface reactivity. Computational and experimental investigations of the gas flow pattern have been performed using the COST jet [43]. In this jet configuration, the electric field is perpendicular to the gas flow and contained between the electrodes, thus, there are no EHD effects expected at the target surface. Figure 4.2 shows experimental (left) and computational (right) investigations of the helium flow pattern when a fused silica disk is introduced as a target at 4 mm (top) and 8 mm (bottom) distance to the nozzle of the COST jet. The experimental investigations were performed with the COST jet (no plasma) with helium gas flow at 1 slm. The helium was entering into anisole atmosphere which was used as a tracer gas to visualize the gas flow with planar laser-induced fluorescence. When the helium effluent hits the surface, a stagnation zone is produced, which laterally directs the flow along the surface [43]. The redirection of the flow and the mixing of helium in the ambient air show distinct differences when comparing the 4 mm and 8 mm distance figures. When the target is placed at 8 mm distance, the redirection of the gas flow leads to a wider lateral spread of the gas. When the target is placed at 4 mm, the redirection of the gas flow also leads to stronger mixing of the effluent with ambient air.

The redirection of the gas flow will impact the species delivery to the surface, as a distance-dependent variation of the lateral spread of the carrier gas helium can be observed. The impact on reactive species delivery to the surface was investigated by

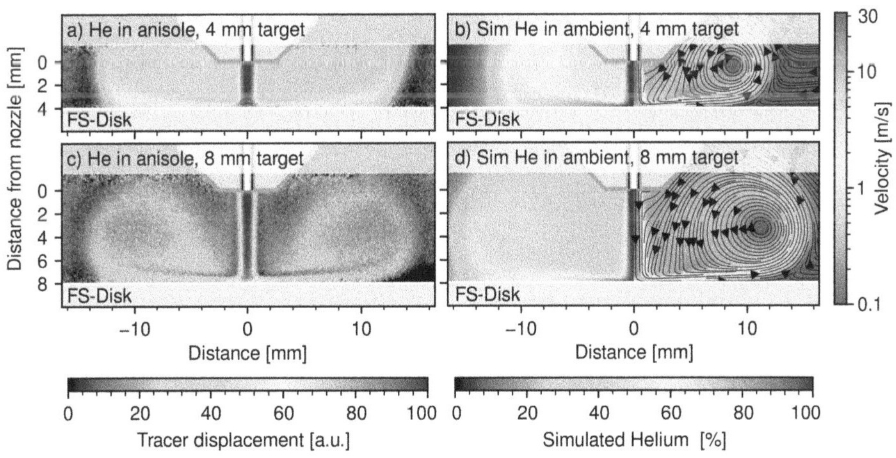

Figure 4.2. Experimental (left) and computational (right) investigation of the flow pattern when a fused silica disk is introduced as target at 4 mm (top) and 8 mm (bottom) distance. Anisole was used as a tracer gas to visualize the gas flow of helium (no plasma) impinging on a solid target at 4 mm (a) and 8 mm (c) distance with planar laser-induced fluorescence. Panels (b) and (d) are simulated flow fields of the helium exiting the jet and impinging on a solid target. The left side in panels (b) and (d) shows the simulated helium mole fraction when entering ambient air, the right side shows the flow velocity. Figure taken from [43], copyright The Authors. Published by IOP Publishing Ltd. CC 4.0.

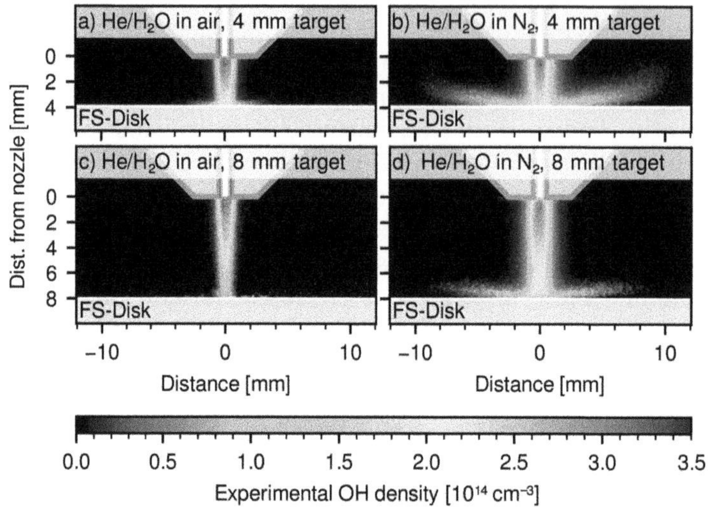

Figure 4.3. OH density measured with planar laser-induced fluorescence in the effluent of the COST jet exiting into ambient air (a, c) and nitrogen (b, d) with a fused silica disk target at 4 mm (a, b) and 8 mm (c, d) distance. Figure taken from [43], copyright The Authors. Published by IOP Publishing Ltd. CC 4.0.

performing planar laser-induced fluorescence of OH when a helium–water (0.25%) gas mixture was used [43]. Figure 4.3 shows the OH densities when the effluent entered ambient air (a, c) or nitrogen (b, d) atmosphere with the fused silica disk placed at 4 mm (a, b) and 8 mm (c, d) distance. The OH density is following the flow pattern of the helium carrier gas in nitrogen atmosphere, where the lifetime of OH is much longer than in air. For both target distances, the lateral spread of OH is significantly greater in nitrogen, showing approximately five-fold of the distance achieved horizontally in air atmosphere. The OH density and distribution changes with target distance and surrounding atmosphere. Note, probing densities (or anything, for that matter) near the interface is rather challenging and comes with uncertainties. For this study, vignetting corrections and careful calibration of the images resulted in a region of uncertainty of $\leqslant 60$ μm, depicted as gray rectangle above the fused silica disk.

Even though the plasma is not in direct contact with the treated target, it can still introduce significant thermal effects at the interface. Depending on the power input, gas composition, gas flow rate, and distance between the plasma source and the substrate, localized heating of the surface can occur due to a combined effect of convective heat transfer, potentially plasma-induced gas flow, and energy deposition from exothermic surface reactions. Chemical reactions between reactive species and surface functional groups can release heat directly at the substrate. For some applications, such as bacteria inactivation or material activation, a moderate temperature rise is beneficial as it can enhance reaction kinetics or promote desired surface transformations. However, in temperature-sensitive processes, such as plasma treatment of biological tissues or polymer surfaces, even small thermal effects must be carefully controlled to avoid damage or unintended modifications. Moreover, plasma-induced heating can affect gas-phase chemistry by altering reaction rates, species lifetimes, and transport dynamics near the surface. Local temperature gradients may also influence boundary layer behavior, leading to enhanced diffusion or changes in flow patterns. Understanding and controlling thermal effects is therefore critical in indirect plasma applications. Techniques such as active cooling of the substrate, pulsed plasma operation, or adjustment of gas flow parameters can be used to mitigate unwanted thermal loads. The spatial thermal dose delivery has also been used to control plasma jets for medical applications [44, 45].

Chapter 4—summary

- Plasma–solid interactions are governed by neutral, charged, photon, and metastable species, each with distinct roles.
- Neutrals enable cleaning, etching, functionalization, and deposition processes such as PECVD and ALD.
- Charged species interact through surface charging, secondary electron emission, and localized discharges.
- UV/VUV photons and metastables deliver energy for ionization, bond breaking, and surface activation.
- Direct plasma sources couple electrically to the target, while indirect sources mainly act through long-lived species, photons, and plasma-induced flow.

Reading recommendations

- Surface functionalization is nicely introduced in 'Foundations of plasma surface functionalization of polymers for industrial and biological applications' [8]. For a deeper dive into surface functionalization, the textbook *Non-Equilibrium Air Plasmas at Atmospheric Pressure* is recommended [2].
- Surface charge dynamics in DBDs are reviewed in [29].
- An extensive review of plasma jets interacting with surfaces is provided in [36].
- Good reading to learn more about plasma actuators and flow control: [41, 42].

References

[1] Hefny M M, Nečas D, Zajíčková L and Benedikt J 2019 The transport and surface reactivity of O atoms during the atmospheric plasma etching of hydrogenated amorphous carbon films *Plasma Sources Sci. Technol.* **28** 035010

[2] Becker K H, Kogelschatz U, Schoenbach K H and Barker R J 2004 *Non-Equilibrium Air Plasmas at Atmospheric Pressure* (CRC Press)

[3] Boselli M, Chiavari C, Colombo V, Gherardi M, Martini C and Rotundo F 2017 Atmospheric pressure non-equilibrium plasma cleaning of 19th century daguerreotypes *Plasma Process Polym.* **14** 1600027

[4] www.relyon-plasma.com/bonding/ Accessed: Aug 2, 2024
[5] Dorai R and Kushner M J 2003 A model for plasma modification of polypropylene using atmospheric pressure discharges *J. Phys. D: Appl. Phys.* **36** 666
[6] Lackmann J -W, Wende K, Verlackt C, Golda J, Volzke J, Kogelheide F, Held J, Bekeschus S, Bogaerts A, Schultz-von der Gathen V and Stapelmann K 2018 Chemical fingerprints of cold physical plasmas–an experimental and computational study using cysteine as tracer compound *Sci. Rep.* **8** 7736
[7] Klinkhammer C, Verlackt C, Śmiłowicz D, Kogelheide F, Bogaerts A, Metzler-Nolte N, Stapelmann K, Havenith M and Lackmann J-W 2017 Elucidation of plasma-induced chemical modifications on glutathione and glutathione disulphide *Sci. Rep.* **7** 13828
[8] Booth J-P, Mozetič M, Nikiforov A and Oehr C 2022 Foundations of plasma surface functionalization of polymers for industrial and biological applications *Plasma Sources Sci. Technol.* **31** 103001
[9] Wolf R A 2012 *Atmospheric Pressure Plasma for Surface Modification* (Wiley)
[10] Banerjee S, Adhikari E K, Sapkota P, Sebastian A and Ptasinska S 2020 Atmospheric pressure plasma deposition of TiO_2: a review *Materials* **13** 2931
[11] Fakhouri H, Salem D B, Carton O, Pulpytel J and Arefi-Khonsari F 2014 Highly efficient photocatalytic TiO_2 coatings deposited by open air atmospheric pressure plasma jet with aerosolized ttip precursor *J. Phys. D: Appl. Phys.* **47** 265301
[12] Fanelli F, Bosso P, Mastrangelo A M and Fracassi F 2016 Thin film deposition at atmospheric pressure using dielectric barrier discharges: advances on three-dimensional porous substrates and functional coatings *Jpn. J. Appl. Phys.* **55** 07LA01
[13] Belmonte T, Henrion G and Gries T 2011 Nonequilibrium atmospheric plasma deposition *J. Therm. Spray Technol.* **20** 744–59
[14] Aghaee M, Verheyen J, Stevens A A E, Kessels W M M and Creatore M 2019 TiO_2 thin film patterns prepared by chemical vapor deposition and atomic layer deposition using an atmospheric pressure microplasma printer *Plasma Process Polym.* **16** 1900127
[15] Verkuijlen R O F, Van Dongen M H A, Stevens A A E, Van Geldrop J and Bernards J P C 2014 Surface modification of polycarbonate and polyethylene napthalate foils by UV-ozone treatment and μ plasma printing *Appl. Surf. Sci.* **290** 381–7
[16] Huiskamp T, Brok W J M, Stevens A A E, Van Heesch E J M and Pemen A J M 2012 Maskless patterning by pulsed-power plasma printing *IEEE Trans. Plasma Sci.* **40** 1913–25
[17] Mameli A, Kuang Y, Aghaee M, Ande C K, Karasulu B, Creatore M, Mackus A J M, Kessels W M M and Roozeboom F 2017 Area-selective atomic layer deposition of In_2O_3: H using a μ -plasma printer for local area activation *Chem. Mater.* **29** 921–5
[18] Schalken J R G, Creatore M, Verhoeven P and Stevens A 2015 μ Plasma printing deposition of amine-containing organo-silane polymers by means of 3-aminopropyl trimethoxysilane *Nanosci. Nanotechnol. Lett.* **7** 62–6
[19] Nijdam S, Teunissen J and Ebert U 2020 The physics of streamer discharge phenomena *Plasma Sources Sci. Technol.* **29** 103001
[20] Jovanović A P, Loffhagen D and Becker M M 2022 Streamer-surface interaction in an atmospheric pressure dielectric barrier discharge in argon *Plasma Sources Sci. Technol.* **31** 04LT02
[21] Babaeva N Y and Kushner M J 2011 Ion energy and angular distributions onto polymer surfaces delivered by dielectric barrier discharge filaments in air: I. Flat surfaces *Plasma Sources Sci. Technol.* **20** 035017

[22] Engelhardt M, Pothiraja R, Kartaschew K, Bibinov N, Havenith M and Awakowicz P 2016 Interaction of an argon plasma jet with a silicon wafer *J. Phys. D: Appl. Phys.* **49** 145201

[23] Yan W, Liu F, Sang C and Wang D 2014 Two-dimensional modeling of the cathode sheath formation during the streamer-cathode interaction *Phys. Plasmas* **21** 013504

[24] Babaeva N Y, Tereshonok D V and Naidis G V 2016 Fluid and hybrid modeling of nanosecond surface discharges: effect of polarity and secondary electrons emission *Plasma Sources Sci. Technol.* **25** 044008

[25] Babaeva N Y and Naidis G V 2020 Ion energies delivered by negative and positive ionization waves to flat dielectric surfaces *Plasma Sources Sci. Technol.* **29** 095020

[26] Viegas P, Slikboer E, Obrusník A, Bonaventura Z, Sobota A, Garcia-Caurel E, Guaitella O and Bourdon A 2018 Investigation of a plasma-target interaction through electric field characterization examining surface and volume charge contributions: modeling and experiment *Plasma Sources Sci. Technol.* **27** 094002

[27] Norberg S A, Johnsen E and Kushner M J 2015 Helium atmospheric pressure plasma jets touching dielectric and metal surfaces *J. Appl. Phys.* **118** 013301

[28] Morsell J K, DeChant C, Gall G S, Trosan D, Lietz A M, Stapelmann K and Shannon S 2024 Estimation of mean electron energy in helium surface ionization waves on dielectric substrates *J. Appl. Phys.* **57** 395202

[29] Komuro A 2025 Recent advances in surface charge dynamics in dielectric barrier discharge: future strategies for control and technological optimisation *J. Phys. D: Appl. Phys.* **58** 133003

[30] Harada Y, Masuda S and Ozaki H 1997 Electron spectroscopy using metastable atoms as probes for solid surfaces *Chem. Rev.* **97** 1897–952

[31] Jánskỳ J, Tholin F, Bonaventura Z and Bourdon A 2010 Simulation of the discharge propagation in a capillary tube in air at atmospheric pressure *J. Phys. D: Appl. Phys.* **43** 395201

[32] Donkó Z, Hamaguchi S and Gans T 2018 The effect of photoemission on nanosecond helium microdischarges at atmospheric pressure *Plasma Sources Sci. Technol.* **27** 054001

[33] Stapelmann K, Gershman S and Miller V 2024 Plasma-liquid interactions in the presence of organic matter–a perspective *J. Appl. Phys.* **135** 160901

[34] Yue Y, Pei X, Gidon D, Wu F, Wu S and Lu X 2018 Investigation of plasma dynamics and spatially varying o and oh concentrations in atmospheric pressure plasma jets impinging on glass, water and metal substrates *Plasma Sources Sci. Technol.* **27** 064001

[35] Morsell J, Trosan D, Stapelmann K and Shannon S 2023 Plasma surface ionization wave interactions with single channels *Plasma Sources Sci. Technol.* **32** 095017

[36] Viegas P, Slikboer E, Bonaventura Z, Guaitella O, Sobota A and Bourdon A 2022 Physics of plasma jets and interaction with surfaces: review on modelling and experiments *Plasma Sources Sci. Technol.* **31** 053001

[37] Viegas P, Hofmans M, Van Rooij O, Obrusník A, Klarenaar B L M, Bonaventura Z, Guaitella O, Sobota A and Bourdon A 2020 Interaction of an atmospheric pressure plasma jet with grounded and floating metallic targets: simulations and experiments *Plasma Sources Sci. Technol.* **29** 095011

[38] Lin L, Hou Z, Yao X, Liu Y, Sirigiri J R, Lee T and Keidar M 2020 Introducing adaptive cold atmospheric plasma: The perspective of adaptive cold plasma cancer treatments based on real-time electrochemical impedance spectroscopy *Phys. Plasmas* **27** 063501

[39] Kogelschatz U 2003 Dielectric-barrier discharges: their history, discharge physics, and industrial applications *Plasma Chem. Plasma Process.* **23** 1–46

[40] Golda J, Held J, Redeker B, Konkowski M, Beijer P, Sobota A, Kroesen G, Braithwaite N S J, Reuter S and Turner M M 2016 Concepts and characteristics of the 'COST reference microplasma jet *J. Phys. D: Appl. Phys.* **49** 084003

[41] Benard N and Moreau E 2014 Electrical and mechanical characteristics of surface AC dielectric barrier discharge plasma actuators applied to airflow control *Exp. Fluids* **55** 1846

[42] Karthikeyan K V and Harish R 2025 Advancements in flow control using plasma actuators: a comprehensive review *Eng. Res. Express* **53** 4244–67

[43] Quesada M J H, Pfaff S, Polito J, Gall G, Kushner M J, Frank J H and Stapelmann K 2025 Two-dimensional mapping of absolute OH densities in an atmospheric pressure plasma effluent via planar laser-induced fluorescence: effects of He/H_2O and He/O_2 mixtures in N_2 and air, with and without solid targets *J. Phys. D: Appl. Phys.* **58** 175205

[44] Gidon D, Graves D B and Mesbah A 2017 Effective dose delivery in atmospheric pressure plasma jets for plasma medicine: A model predictive control approach *Plasma Sources Sci. Technol.* **26** 085005

[45] Gidon D, Graves D B and Mesbah A 2019 Spatial thermal dose delivery in atmospheric pressure plasma jets *Plasma Sources Sci. Technol.* **28** 025006

Chapter 5

Plasma–liquid interactions

Plasma–liquid interactions tie together plasma physics, interfacial transport, and solution chemistry across different time and length scales. Unlike plasma–solid systems, the boundary here can move, reshape, and react: liquids charge and screen applied fields, surfaces deform under electric and shear stresses, and reactive fluxes are continuously redistributed by adsorption, dissolution, reaction, and flow. The result is a coupled problem in which electrical properties, gas composition and flow, liquid formulation, and geometry jointly determine where energy and matter enter the system and what chemistry follows.

This chapter builds a practical map for that coupling. We first outline the major discharge types encountered with discharges in contact with a liquid, discharges inside liquids, and multiphase cases involving bubbles, droplets, aerosols, or foams, highlighting how geometry and timescales bias ignition, transport, and heating (section 5.1). We then turn to neutral-driven chemistry and transport, from interfacial adsorption to bulk uptake and reaction–diffusion behavior, with a critical look at boundary conditions and transport theories used in models (section 5.2). Next, we examine charged species at the interface, where interfacial fields, double layers, solvated electrons, and Faradaic pathways tie plasma dynamics to electrochemistry (section 5.3). Photon and metastable effects are treated as predominantly interfacial drivers that deposit energy and trigger photolysis or Penning-type processes with short penetration depths (section 5.4). Finally, we analyze plasma-induced fluid motion and interfacial dynamics, showing how Maxwell stress, ion momentum, Marangoni forces, and gas-jet shear reshape the surface and feed back on transport and chemistry (section 5.5). The chapter closes by contrasting direct versus indirect sources, clarifying when electrical coupling to the liquid dominates and when neutrals, photons, and flow carry the dose (section 5.6).

doi:10.1088/978-0-7503-6254-2ch5

5.1 Concepts of plasma–liquid discharges

A variety of setups and geometries can be found in the literature for plasma–liquid interactions. Most of these systems can be sorted into one of the following categories [1]:

- Discharge in the gas phase in contact with a liquid: with and without electrical coupling, at the plasma–liquid interface as surface discharge,
- Discharge inside the liquid,
- Multiphase discharges: inside bubbles in liquids or discharges contacting droplets, aerosols, or foams.

Each of these plasma–liquid systems has unique interactions, with differences in plasma generation, heat and mass transfer, and species transport. The different concepts will be introduced briefly in the following. For further reading, reviews and roadmap publications are recommended [1–3].

At-a-glance

Plasma–liquid discharges can occur in the gas phase in contact with a liquid, inside the liquid, or in multiphase systems. Geometry and liquid properties strongly shape discharge initiation, heat transfer, and species production.

- Gas-phase discharges near liquids: Pin-to-water, jets, and DBDs generate species above the liquid, with surface instabilities and ozone chemistry often playing key roles.
- Discharges inside liquids: Breakdown requires extreme fields; cavitation or bubbles enable nanosecond streamer initiation and lead to unique species formation compared with gas-phase systems.
- Multiphase discharges: Bubbles, aerosols, and foams provide microreactors for localized plasma chemistry, bridging gas–liquid complexity.

5.1.1 Discharge in the gas phase in contact with a liquid

Various gas-phase discharges near or at liquid surfaces have been widely studied, especially configurations involving a metal pin and a water electrode. Water electrodes tend to stabilize discharges, preventing constriction, likely due to their resistivity and possibly other factors. Under typical DC conditions (gaps of a few mm, currents of 5–50 mA, voltages 1 kV), diffuse glow discharges can be sustained in atmospheric pressure air. When the liquid acts as the cathode, heat transfer is more efficient, leading to significant evaporation. Reported gas temperatures exceed 3000 K with electron densities around 10^{19} m^{-3}. In some cases, self-organized patterns form at the liquid surface, particularly when the liquid is the anode (more in section 5.5). High electric fields near the liquid may cause plasma–liquid interface instabilities, potentially enhancing species transfer. Pin-to-water setups can also generate corona or spark discharges, with corona discharges producing ionic wind that deforms the liquid surface (more in section 5.5). These have been less thoroughly explored, though. Discharges that propagate along the gas–liquid interface, e.g., surface streamers, have also been studied. The ionization waves travel at 1–10 km s^{-1} and may promote enhanced transfer of reactive species directly at the interface.

Plasma jets represent a major research focus in plasma–liquid interaction studies. Not all contact the liquid electrically; instead, plasma-generated species are delivered by the gas flow. Convective transport enhances reactive species transfer into the liquid. A key distinction is whether the electric field aligns parallel (linear jet) or perpendicular (cross-field jet) to the gas flow. Cross-field jets usually don't touch the liquid but can produce guided streamers that propagate ionization fronts toward the liquid. When atmospheric pressure plasma jets (APPJs) do not contact the liquid, the reactive species interacting with the liquid are mainly neutral species, with (V)UV photolysis also contributing. Comparative studies show significant differences in liquid-phase chemistry between contacting and non-contacting plasma configurations, a finding confirmed both by models and initial experiments. For APPJs operating with oxygen-containing gases, ozone often plays a central role in gas-phase reactivity.

Dielectric barrier discharges (DBDs) are widely used for atmospheric pressure plasma applications but have been mostly studied with solid electrodes. In DBDs in contact with liquids, a dielectric barrier (such as glass or ceramic) separates the electrode from the discharge zone, but the liquid itself can act as a counter electrode. When the plasma is generated adjacent to the liquid, reactive species produced in the gas phase diffuse into the liquid. The presence of the electric field can cause electrohydrodynamic (EHD) effects at the liquid surface.

5.1.2 Discharge inside the liquid

Discharges inside liquids require a rapid breakdown process and very high voltages due to the high density of the liquid. Non-thermal plasmas in liquids are typically streamer or corona discharges in pin-to-pin or pin-to-plate configurations. In liquid discharges, current continuity is often achieved through slow ions and displacement current, rather than mobile electrons like in gas phase discharges. The breakdown mechanism depends on the liquid properties and the applied voltage. Under certain conditions, the discharge is initiated by cavitation [4–8]. Cavitation depends strongly on the electrode tip radius and the applied voltage rise time. Experimental [4, 5] and theoretical [6–8] work could show that cavities develop for eletrode tip radii $\leqslant 30\ \mu m$ and a voltage rise time of <10 kV ns^{-1}. This causes a negative electrostrictive pressure $|P_-|$ in the range of 10s of MPa:

$$|P_-| = \frac{1}{2}\alpha\varepsilon_0\varepsilon E^2 \tag{5.1}$$

with α an empirical factor for a dielectric liquid ($\alpha = 1.5$ for water), and ε the dielectric constant of the liquid ($\varepsilon = 81$ for water), and E the electric field. It would require 100s of nanoseconds for the water to move in and compensate the negative pressure in the void. Nanocavities can grow to enable ionization and discharge initiation on nanosecond timescales as the water does not have time to respond. In order to ionize water molecules, an energy of $E \geqslant 12.6\,\text{eV}$ is required. Thus, the cavity needs to be sufficiently large to accelerate seed electrons to this energy. The speed of expansion can be estimated as:

$$v_{\exp} \approx \sqrt{\frac{2|P_{-}|}{3\rho}} \tag{5.2}$$

with ρ the density of the liquid. This assumes that the negative pressure in the cavity is sufficient to move water molecules. For a seed electron to be accelerated to 12.6 eV in the electric field E in the cavity, the cavity size $R(t)$ should be such that:

$$12.6\ \text{eV} = e\Delta V = 2eER(t) \approx 2eE\sqrt{\frac{2|P_{-}|}{3\rho}}\,t_{\text{ion}} \tag{5.3}$$

with t_{ion} the time needed for a cavity to expand sufficiently such that the electron can gain 12.6 eV. The rise time of the voltage must be shorter than the time required for new water to fill in the region close to the electrodes where voids can grow for cavitation to occur.

If the timescales do not allow for cavitation, several other mechanisms can lead to breakdown in liquid. DC or AC excitation can lead to Joule heating and produce a bubble through which breakdown can occur [3]. Similarly, pre-existing bubbles in the liquid can facilitate breakdown [9].

Discharges in liquid have high electron densities that can reach 10^{24}–10^{26} m^{-3} with ionization degrees of 1–10^{-3} [1]. Gas temperatures of 3000–5000 K can be achieved. The discharge channels are rather small with diameters $\approx$10 μm, leading to power densities of 10^{15} W m^{-3} and more. The electric field in the propagating streamer heads can be 1 MV cm^{-1} and more.

The reactive species production in discharges in liquids can be quite different from the reactive species production in gas phase discharges. If water is the liquid, electrons will dissociate water, forming H + OH + e. Then, recombination reactions will lead to the formation of more stable species, such as H_2O_2, H_2 etc. A detailed description of the chemical reaction network is provided in [10, 11].

5.1.3 Multiphase discharges

Multiphase discharges can be discharges inside bubbles in liquids or discharges contacting droplets, aerosols, or foams. The disparate permittivity, conductivity, and curvature across either bubble-liquid or gas-droplet systems partition the applied voltage and create preferential sites for breakdown. Both have in common that the surface-area-to-volume ratio is optimized to overcome transport limitations (more in section 5.2.1).

Distinct ignition routes can be observed in bubbles in liquids depending on the timescale of the voltage pulse. Experiments with argon bubbles in de-ionized water have shown that for nanosecond pulse excitation (~10 ns FWHM), breakdown typically initiates at the sharpened anode tip through cavitation and rapidly travels towards and into the bubble [4]. The voltage pulse is too short for appreciable charging of the interface and the discharge results in a partial Townsend discharge inside the bubble [4]. On sub-microsecond timescales (~500 ns), a direct breakdown in the bubble volume is less common. Ionization starts at the tip and spreads along the bubble surface. After the applied voltage terminates, a delayed re-illumination consistent with a DBD return-stroke current appears. Interfacial charging during the pulse partially decouples the field and then drives a short post-pulse discharge [12]. On microsecond timescales, Ohmic heating can vaporize water at the electrode tip, leading to vapor expansion, local field reshaping, and breakdown occurring either at the electrode tip or directly within the bubble. In perturbed bubbles, the streamer can show a hybrid of volume and surface propagation [12].

The conductivity of the liquid impacts the balance of displacement and free current, screens interfacial fields more quickly, and limits how much voltage can 'reach' the gas bubble [12, 13]. It is convenient to compare the timescales of the voltage pulse, in particular the rise time t_{rise} and pulse length t_{pulse} to the Maxwell relaxation time of the liquid $\tau_M = \varepsilon/\sigma$:

- $t_{\text{rise}}, t_{\text{pulse}} \ll \tau_M$ (poorly conducting liquid, e.g., DI water with $\tau_M \approx 5\ \mu\text{s}$): the interface cannot shed polarization/space charge during the pulse. The field is set mainly by instantaneous geometry and permittivity contrast. The field enhancement at the electrode dominates ignition and return-stroke behavior is suppressed.
- $t_{\text{rise}} \approx \tau_M$ (low to moderate σ, 10s–100s of ns): charge builds up on the interface, enabling post-pulse discharge in a return stroke and promoting surface-guided ionization along the gas–liquid interface.
- $t_{\text{pulse}} \gg \tau_M$ (moderate to high σ): liquid screens rapidly, Joule heating and vapor growth reconfigure the interface. Thermally assisted breakdown channels and hybrid surface and volume streamers in the bubble.

Droplets and aerosols are the complementary case: gas outside, liquid phase inside, often with extreme curvatures and large surface areas. The same permittivity contrast as for bubbles applies. The steep permittivity gradient focuses field lines, e.g., at droplet edges or plateau borders in foams. Small droplets ($\lesssim 10\ \mu\text{m}$) of moderate permittivity are enveloped by the streamer and do not perturb streamer propagation, while larger droplets with a big enough capacitance can intercept, re-initiate, or stall the streamer [14]. Microdroplets in an RF glow discharge accelerate in the discharge due to differential charging and sheaths bounding the droplet [15]. Droplet evaporation can lead to strong inhomogeneities in the plasma and localized increased OH densities around the droplets [16]. The size of the droplets, and for some species Henry's law constant, significantly influence the solvation dynamics and concentration of reactive species in the droplets. Larger droplets may deplete high Henry's law gas-phase reactive species before reaching liquid-phase saturation, limiting in-liquid density. Species with low Henry's law constant come into equilibrium more rapidly, often without depleting local gas phase species; thus, the dependence on droplet diameter is weaker [17]. Plasma–aerosol interactions have recently been reviewed in [14, 15].

5.2 Neutral species at the interface and transport

At-a-glance

Neutral atoms and molecules dominate plasma–liquid chemistry, with adsorption, solubility, and residence time determining whether they react at the surface or penetrate into the bulk.

- Transport theory: Multiple models (Henry's law, kinetic, diffusive, film theories) attempt to describe uptake, but their accuracy depends on solubility and non-equilibrium effects.
- Reaction–diffusion dynamics: Highly reactive species like OH remain near the interface, while more stable species (e.g., H_2O_2) diffuse deeper, producing distinct spatial reactivity profiles.

Neutral species, including radicals, reactive molecules, and metastable atoms (cf section 5.4), play a critical role in plasma–liquid interactions. Unlike charged particles, they are not influenced by electric fields, but they are key agents in initiating chemical transformations in the liquid phase. Figure 5.1 shows a schematic depiction of neutrals approaching the liquid interface and their interaction mechanisms.

When neutrals approach the liquid surface, they may first adsorb before reacting or dissolving. Adsorption efficiency depends on molecular polarity, surface tension effects, and the presence of surface-active agents such as surfactants. Adsorption can alter surface properties, potentially affecting surface tension and interface

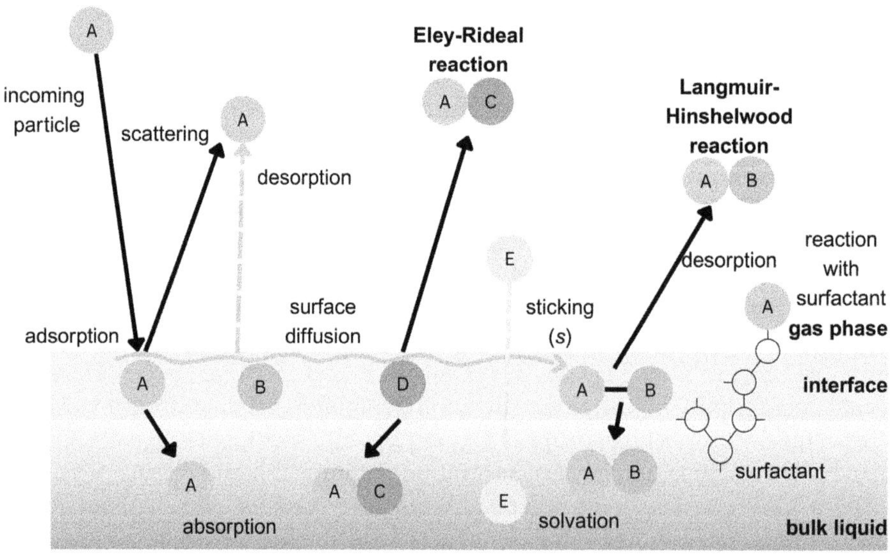

Figure 5.1. Neutral species approaching a liquid interface.

composition. Surface reactions of neutrals can follow mechanisms analogous to those on solids (cf section 2.2). The two surface reaction mechanisms Langmuir–Hinshelwood and Eley–Rideal can be applied to a liquid interface as well [1]. In the Langmuir–Hinshelwood mechanism, two species (A and B) adsorb and react on the surface, producing AB, which may desorb or dissolve depending on its solubility. In contrast, the Eley–Rideal mechanism involves a gas-phase species (A) reacting directly with an adsorbed species (C), producing AC, which may either desorb and return to the gas phase or dissolve, depending on its solubility. Given the high neutral flux in plasma environments, the Eley–Rideal pathway may be significant. Some neutral species, like OH radicals or ozone, may also react immediately at the interface with components of the liquid, such as dissolved organic compounds or surfactants. Organic layers, contaminants, or surfactants can then hinder neutral species transport or selectively react with them, which in turn alters the interfacial chemistry. Neutrals can also be scattered. The result is the same as for desorption, the species will go back into the gas phase. However, if the species does not reside on the water surface long enough to reach thermal equilibrium with the water, i.e., there is no significant energy exchange, then this process is defined as scattering [19].

Recent molecular dynamics (MD) simulation studies have grouped plasma-relevant species into two categories based on their residence time on the water surface (given for 100 ps) [19]:

- short-residence species (O_3, N_2O, NO, NO_2): desorb within 100 ps with ≈95% probability. Scattering as the second-most probable process; absorption is rare (⩽0.5%);
- long-residence species (OH, H_2O_2, HNO_3, and N_2O_5): show high adsorption probability (≈78% for N_2O_5 and H_2O_2, to 90% for HNO_3), multiple absorption and adsorption events before desorption occurs, comparably large desorption probabilities (10%–20%).

The surface population of species at the plasma–water interface can be expressed as an interface population balance [19]:

$$\frac{\partial n_s}{\partial t} = \frac{1}{4} s \cdot n_g v_{th} - k_{des} n_s - k_{abs} n_s \tag{5.4}$$

with n_s the surface density of the species, s its sticking coefficient (defined as 1 minus the scattering probability), n_g the species density in the gas phase, and v_{th} the species' thermal velocity. The first term of equation (5.4) is the flux of species from the plasma towards the interface. The second term balances the flux of incoming species with the desorption of species with the rate constant k_{des}, the third term accounts for species getting absorbed by the liquid with the rate constant k_{abs}. Reference [19] provides tables for kinetic rate constants, characteristic residence times, and sticking coefficients for different water temperatures for the above-mentioned nine species.

Once at the interface, soluble species can enter the liquid bulk directly, depicted with species E. A rough or disturbed interface (e.g., due to gas flow or due to a flowing liquid, as we will explore in section 5.5) increases the effective surface area, potentially enhancing neutral species uptake. In stirred systems or, more general, under continuous renewal of the interface, the interface exposes fresh liquid, enhancing mass transfer and preventing saturation effects. The uptake of the species further depends on the species' solubility and the transport theory applied.

5.2.1 Transport theory

Theory to describe the transport across the interface in plasma–liquid systems is still under debate. A recent publication by Heirman and Bogaerts provides a critical comparison of interfacial boundary conditions and advocates for using film theory in combination with the mass accommodation coefficient [20].

Table 5.1 summarizes the transport models at plasma–liquid interfaces that will be introduced in this section and gives their key assumptions, strengths and limitations.

The simplest case, following Henry's law, has been used in several computational publications. Here, the uptake depends on the solubility of the species and is described by Henry's law constants H. Henry's law relates the solubility via concentration c of a species to the partial pressure of the gas P_{gas}:

$$c = H_{\mathrm{S}}^{\mathrm{cp}} \cdot P_{\mathrm{gas}} \tag{5.5}$$

At constant temperature and equilibrium, the density of a gas dissolved in a liquid is proportional to the partial pressure of the gas in the vapor phase. Another way to define Henry's law constant is by using the dimensionless ratio between liquid species concentration and gas phase concentration:

$$H_{\mathrm{S}}^{\mathrm{cc}} = \frac{c_{\mathrm{l}}}{c_{\mathrm{g}}} \tag{5.6}$$

Larger Henry's law constants mean a greater likelihood that the species in contact with the water will dissolve into the liquid. Note, Henry's law is not applicable if there is a chemical reaction occurring between solute and solvent. Table 5.2 provides Henry's law solubility constants $H_{\mathrm{S}}^{\mathrm{cp}}$ for selected species relevant for plasma–liquid interactions [21]. A compilation of more than 46 000 Henry's law constants for 10 173 organic and inorganic species is also available on www.henrys-law.org.

Henry's law is widely used in plasma–liquid modeling, but its assumptions rarely hold under non-equilibrium, transient conditions. It requires interfacial equilibrium and purely diffusive transport, whereas atmospheric pressure discharges can drive additional fluxes through EHD motion, electric surface stresses, and gas-jet shear (see sections 5.5 and 5.6). Consistent with experiments, Henry-law boundary conditions perform reasonably for highly soluble species such as H_2O_2 [20], but they overestimate uptake of species with intermediate or low solubility (e.g., O_3, NO, HNO_2). The error arises because enforcing equilibrium at each time step implies instantaneous dissolution; this is acceptable for highly soluble species but not for others, where finite interfacial kinetics and renewal times control the approach to equilibrium.

Table 5.1. Comparison of transport models at plasma–liquid interfaces.

Model	Key assumption	Main parameter(s)	Strengths	Limitations
Henry's law	Equilibrium at the interface	Henry's law constant H^{cc}	Simple; valid for highly soluble species	Overestimates low-solubility species; ignores non-equilibrium
Kinetic model	Steady-state flux exchange	Mean velocity $\overline{v}$, H^{cc}	Captures dynamic uptake; suitable for transient systems	Requires species-specific data; idealized assumptions
Diffusive model	Weighted diffusion between phases	Diffusion coefficients D_g, D_l	Includes diffusion control	Requires spatial modeling; equilibrium shuts off flux
One-film theory	Stagnant liquid film as barrier	Film thickness z_f, $D_{i,w}$	Simple resistance model; accounts for diffusion time	Only liquid-side barrier; simplified interface geometry
Two-film theory	Stagnant films on both sides	Mass transfer coefficients $k_{i,l}$, $k_{i,g}$	Accounts for both gas and liquid barriers	Requires k_i data; increased complexity
Film theory + mass accommodation	Combined resistance and uptake probability	α_b, D_l, Δx	Flexible; fits transient and reactive systems	Parameter-intensive; depends on interface geometry

Table 5.2. Henry's law solubility constants for selected species relevant for plasma–liquid interactions. All constants from [21], except for atomic oxygen O [22, 23].

Species	Henry's law constant H_S^{cp} mol m^{-3} Pa^{-1}	Dimensionless Henry's law constant H_S^{cc}
O_2	1.3×10^{-5}	3×10^{-2}
O	$\approx 4 \times 10^{-3}$ [22], $\leqslant 8 \times 10^{-3}$ [23]	≈ 10 [22], $\leqslant 20$ [23]
N_2	6.5×10^{-6}	1.6×10^{-2}
H_2O_2	8.6×10^{2}	2.13×10^{7}
OH	3.8×10^{-1}	9.42×10^{2}
HO_2	6.8	1.69×10^{4}
NO	1.9×10^{-5}	4.7×10^{-2}
NO_2/HNO_2	$1.2 \times 10^{-4}/4.8 \times 10^{-1}$	$2.97 \times 10^{-1}/1.19 \times 10^{3}$
NO_3/HNO_3	$3.8 \times 10^{-4}/2.1 \times 10^{3}$	$9.42 \times 10^{-1}/5.2 \times 10^{6}$

Several different boundary conditions have been used over the past decade, also summarized and reviewed in [20]. In brief, the different approaches can be grouped into either kinetic boundary conditions, diffusive boundary conditions, or film theory. The kinetic boundary condition introduces a flux between gas and liquid phase, as well as between liquid and gas phase, taking into account the not necessarily equilibrium steady-state for plasma–liquid interactions [20, 24]:

$$\Gamma_{g\leftrightarrow l} = \frac{1}{4} \cdot \overline{v}\left(c_g - \frac{c_l}{H^{cc}}\right) \tag{5.7}$$

For diffusive transport from the gas into the liquid phase, a weighted diffusion coefficient is introduced [20, 25]:

$$D_{gl} = D_g\left(\frac{H_S^{cc} n_g - n_l}{H_S^{cc} n_g}\right) \tag{5.8}$$

with D the diffusion coefficient (subscript gl: diffusion between gas and liquid, subscript g: diffusion in the gas phase), $n_{g,l}$ densities in the gas phase (g) and liquid phase (l), respectively. When equilibrium density at the surface is reached, diffusion is shut off.

Mass transfer at interfaces is described more generally using the film theory, which assumes the presence of a stagnant liquid layer adjacent to the interface. This stagnant layer acts as a reservoir, temporarily absorbing species before they diffuse into the bulk liquid. In essence, it introduces a resistance to the flux of gaseous species. Species accumulate in this layer until equilibrium with the gas phase is reached, preventing further uptake unless diffusion into the bulk reduces their concentration. The one-film theory expresses the flux as [20, 26]:

$$\Gamma_{g\leftrightarrow l} = \frac{D_l}{\Delta x}(H^{cc} \cdot c_g - c_l) \tag{5.9}$$

where Δx is the film thickness, in [26] assumed to be $\Delta x \approx 10\ \mu m$, and in [27] $\Delta x = 1$ mm, based on environmental studies they reference in their work. The factor $\frac{D_l}{\Delta x}$ essentially introduces a resistance to the flux, caused by the stagnant layer. Equation (5.9) describes the flux towards the liquid when the liquid-side is the driving force. This is usually preferred when the gas phase is well mixed and the resistance is mainly on the liquid side. When the gas-side is the limiting step, using equation (5.6), the flux is rearranged to account for the different driving forces, as used by Silsby *et al* [27]:

$$\Gamma_{g\leftrightarrow l} = \frac{D_l}{\Delta x}\left(c_g - \frac{c_l}{H^{cc}}\right) \tag{5.10}$$

The two-film theory, commonly applied in environmental and aerosol science, expands on this concept by adding a stagnant film on the gas side of the interface. This second film hinders the transport of hydrophilic species into the liquid and likewise slows the desorption of hydrophobic species from the liquid into the gas phase. Recent work by Silsby *et al* indicates that the two-film model better describes mass transfer across gas–liquid interfaces [27]:

$$\Gamma_{g\leftrightarrow l} = (H^{cc}k_l + k_g)\cdot\left(c_g - \frac{c_l}{H^{cc}}\right) \tag{5.11}$$

where k_l and k_g are the mass transfer coefficients of the species due to the stagnant film on the liquid and gas sides of the interphase. Mass transfer coefficients are not readily available for all species relevant for plasma–liquid interactions. The model introduces added complexity, notably the need for a Sherwood number, a dimensionless quantity expressing the ratio of convective mass transfer to diffusion. Experimental data for plasma-generated species are scarce due to their high reactivity, making parameter validation challenging.

In [20], a similar but different approach is proposed. Instead of using mass transfer coefficients k_i, bulk mass accommodation coefficients α_b are incorporated. Mass transfer coefficients represent the overall rate of mass transfer between the two phases gas and liquid, while the mass accommodation coefficient represents the probability of a gas molecule being dissolved into a liquid. The flux equation then becomes [20]:

$$\Gamma_{g\leftrightarrow l} = \alpha_b\cdot\frac{D_l}{\Delta x}(H^{cc}\cdot c_g - \bar{c}_l) \tag{5.12}$$

The maximum film thickness is given by $\Delta x = \sqrt{2\cdot D_l\cdot r/\overline{v}_{int}}$ with r, the radial dimension of the gas–liquid interface, and the average velocity at the interface $\overline{v}_{int}$. The liquid concentration is given as an average concentration $\bar{c}_l$, as it is difficult to define a bulk concentration with constant transport through convection in their model [20].

While applying film theory in conjunction with mass accommodation or mass transfer coefficients helps to describe mass transport more accurately, significant uncertainties remain, especially under non-equilibrium conditions typical of plasma–liquid systems. For instance, evaporation processes are still poorly understood at the nanometer scale and on sub-nanosecond timescales. Additionally, plasma-specific phenomena such as chemical reactivity, electric fields, currents, EHD forces, and the formation of a plasma sheath further complicate the dynamics across vastly different time and length scales. As plasma-induced reactions proceed, the equilibrium concentrations in both the gas and liquid phases shift, dynamically influencing interfacial transport in both directions.

Experimental studies face limitations due to the lack of diagnostics capable of probing these ultrafast and ultrasmall interfacial phenomena. One possible approach to overcome this is combining experimental studies at accessible scales with advanced models capable of predicting behavior across those scales, potentially improving our understanding of interfacial transport and plasma-induced effects.

5.2.2 Reaction–diffusion dynamics

For highly reactive species (e.g., OH radicals, atomic oxygen), reaction rates may be fast enough that they are consumed at or near the interface, leading to concentration gradients confined to nanometer–micrometer scales. The knowledge about the interface is mostly from theory, experiments are difficult to perform close to the interface. One recent investigation published by Poggemann *et al* [28] is combining MD simulations, experiments, and kinetic modeling to bridge the gap between the vastly different time and length scales that are accessible through MD simulations and experiments. They used the one-film theory as described by equation (5.9) and [26] to implement mass transport and established a reaction–diffusion model to describe transport of species. The study was performed with a focus on OH and H_2O_2 with the one-dimensional reaction–diffusion equation [28]:

$$\frac{\mathrm{d}n_{i,1}}{\mathrm{d}t} = R_i + D_{i,1}\frac{\mathrm{d}^2 n_{i,1}}{\mathrm{d}z^2} \tag{5.13}$$

with R_i the reaction term, including production and consumption reactions of species i in the liquid. The experimental and computational results have shown that the highly reactive OH is mainly located at the surface of the liquid (less than $\approx 1\ \mu$m penetration depth within 200 s) with good agreement between experimental observations and model predictions. H_2O_2's transport is mainly diffusion-driven with a penetration depth of $\approx$1 cm at 200 s. However, the diffusion coefficients obtained by MD simulations and experimentally differ, leading to discrepancies in the model-predicted transport behavior. This can be explained by the MD simulation only taking into account diffusion, while the experiment was performed with a plasma jet that could introduce (forced and free) convective transport as well. Furthermore, gravity might play a role. This motivated the use of an effective diffusion coefficient, determined experimentally, that takes into account all contributions to transport occurring in a real system [28].

Previous MD simulations have shown on small length (0.8 nm) and time (100 ps) scales that OH radicals do penetrate into water, react with water, and exchange a hydrogen atom, forming a new OH radical and water molecule [29]. HO_2, on the other hand, remains at the interface and is not penetrating into the bulk liquid, as it is energetically favorable to bind to the outside of a water cluster. HO_2 radicals can react with water to form superoxide anions O_2^-, or with the hydronium ion H_3O^+ to form H_2O_2 and O_2. H_2O_2 was only found to react when 3 H_2O molecules were sufficiently close, forming 2 HO_2 and 2 H_2O [29]. An experimental study used two-photon absorption laser-induced fluorescence in the liquid to investigate the solvation of atomic oxygen into the liquid [22]. Solvated O was detectable up to a depth of 0.4 mm. It should be noted that this experiment also used a plasma jet where transport is not only diffusion-based. For diffusive transport only, using the diffusion coefficient for O in water, a penetration depth of 0.5 μm was estimated, well below the resolution of most experimental setups at this time [22]. Similarly, theoretical work by Rumbach *et al* suggests an upper limit of penetration depth for OH radicals on the order of 100 nm [30]. An overview of selected diffusion coefficients relevant for plasma–liquid interactions is provided in table 5.3.

Table 5.3. Selection of reported diffusion coefficients in water at 300 K relevant for plasma–liquid interactions.

Species	Diffusion coefficient/$m^2\ s^{-1}$	Note	Reference
H_2O_2	$(6 \pm 0.9) \times 10^{-10}$	Water, reactive MD	[28]
	1.34×10^{-9}	Water, reactive MD	[29]
OH	1.1×10^{-9}	Water, reactive MD	[28]
	8.37×10^{-9}	Water, reactive MD	[29]
	0.7×10^{-9}	Water, no reactivity	[31]
NO	4.5×10^{-9}	Buffer solution, experimental	[32]
	2.07×10^{-9}	Water, experimental	[33]
	$(3.07 \pm 0.47) \times 10^{-9}$	Water, no reactivity	[34]
HO_2	6.8×10^{-10}	Reactive MD	[29]
H_2O	$(1.2 \pm 0.026) \times 10^{-9}$	Reactive MD	[28]
	2.31×10^{-9}	Experimental	[35]
H	1.2×10^{-9}	Reactive MD	[29]
O	1.2×10^{-9}	Reactive MD	[29]
	1.5×10^{-9}	Reactive MD	[22]

5.3 Charged species at the interface and transport

At-a-glance

Charged particles interact with structured interfacial fields, forming double layers, solvated electrons, and driving electrochemical (Faradaic) reactions that tightly couple plasma and liquid electrochemistry.

- Electrostatic double layer: The plasma sheath couples with the Debye layer in the liquid, with ion distributions and potentials governed by ionic strength and relaxation times.
- Secondary electron emission: Rare but possible at liquid interfaces under ion bombardment or strong fields, adding electrons back to the plasma and influencing local chemistry.
- Faradaic processes: Plasma-supplied electrons or ions drive reduction and oxidation at the interface, enabling plasma-assisted electrochemistry and material synthesis.

As was described in sections 3.1.3 and 3.2.2, charged species will only reach the interface when a direct plasma source is used. Charged species (ions and electrons) encounter a highly structured environment at the plasma–liquid interface, shaped by electric fields, interfacial charge layers, and electrostatic interactions. Their behavior influences charge transfer, surface reactions, and EHD phenomena.

The integration of surface charge dynamics, field enhancement due to dielectric boundaries, and mass transport are fundamentally dependent on the dielectric relaxation time of the liquid, or in other words, how the liquid is able to accumulate and hold charge at the surface. The liquid can be approached as a dielectric which becomes more and more 'lossy' when ions are introduced. From an electrical standpoint, the most important parameters of a liquid are its permittivity ε and its electric conductivity σ. Ions effectively neutralize space charge in the dielectric, macroscopically quantified with the Maxwell relaxation time τ_M, derived from the (free) charge continuity equation with $\partial\rho_f$ the free charge density:

$$\frac{\partial \rho_f}{\partial t} + \nabla \cdot \mathbf{J_f} = 0 \tag{5.14}$$

$$\frac{\partial \rho_f}{\partial t} + \frac{\sigma}{\varepsilon}\rho_f = 0 \tag{5.15}$$

such that the solution is:

$$\rho_f(t) = \rho_0 \cdot \exp\left(-\frac{t}{\tau_M}\right) \tag{5.16}$$

and $\tau_M = \varepsilon/\sigma$. In plain words, the Maxwell relaxation time is related to the time it takes for charges to redistribute in a dielectric material in response to an external electric field. Maxwell and Debye relaxation are both related to the polarization behavior of dielectric materials, but based on different physical mechanisms. The Debye relaxation time describes the time it takes polar molecules to reorient the dipole moment in response to an external electric field, originally described by Debye in 1929 [36]. For water, $\tau_D \approx$ 8 ps [37] and $\tau_M \approx$ 5 μs (for de-ionized water with $\sigma = 1.5$ μs cm^{-1}).

In models it is typically assumed that ions just enter the liquid as their energy can overcome any barriers but the validity of this assumption is difficult, if not impossible, to prove [20]. Ions, depending on the liquid used, can also recombine with counter-charges at the interface or be neutralized by electrons at the interface before entering the liquid. Electrons may enter the liquid, leading to solvated electrons. In positively biased plasma sources, gaseous ions flow towards the interface and can ionize water to produce solvated, or hydrated, electrons [38], although this hypothesis is difficult to prove and subject to current debate and ongoing research. In negatively biased plasma sources, electrons are injected directly. Regardless of origin, solvated electrons are extremely reactive and with a reduction potential of –2.88 V versus NHE (normal hydrogen electrode) in water, one of the most powerful reducing agents [38]. However, solvated electrons have small penetration depth, an upper bound estimated as $\approx$ 100 nm [38] and experimentally reported values of 12.3 $\pm$ 4.9 nm [39]. In the absence of scavengers, the interfacial concentration $[e^-_{aq}]_0$ and penetration depth l scale with the current density of the plasma source [38]: $[e^-_{aq}]_0 \sim j_e^{2/3}$ and $l \sim j_e^{-1/3}$; are scavengers present and abundant: $[e^-_{aq}]_0 \sim j_e$ and l becomes independent of the current density as the penetration depth now completely depends on the scavenger concentration.

Solvated electrons quickly self-recombine [40]:

$$2\,e^-_{aq} + 2\,H_2O \rightarrow 2\,OH^-_{aq} + H_{2\,aq} \tag{5.17}$$

or react with dissolved species, such as a metal cation:

$$e^-_{aq} + S^n_{aq} \rightarrow S^{n-1}_{aq} \tag{5.18}$$

In neat water, the local electron concentration depends on their diffusion into the water and elimination by reaction (5.17). The higher the electron concentration (higher current density j_e), the quicker electrons react with each other via equation (5.17), the smaller the penetration depth, explaining the inverse relationship $l \sim j_e^{-1/3}$.

With the water or electrolyte solution being exposed to high electric fields, the orientation of water molecules is affected. Polarization density increases as a function of electric field strength [41]. The water molecules orient themselves along the electric field direction to minimize dipole-field interaction energy. But water has another means to minimize the interaction energy, by increasing the O–H distance or by reducing the (H–O–H) angle [41]. This can then lead to an increased evaporation when the electric field is perpendicular, or to impeded evaporation when the electric field is parallel [42]. MD simulations found that the electric field-induced orientation of the water molecules accelerates sputtering of water molecules by impinging ions, when the electric field is larger than 10^9 V m^{-1} [43].

When charged species are approaching an interface, a sheath forms. Not much is known about the sheath at a plasma–liquid-interface at this point, as pointed out in the perspective paper by Vanraes and Bogaerts in 2021 [44]. A few studies have been performed, of theoretical, computational, and less of experimental nature. In the following subsections, some studies involving the formation of electric double layers and some highlighted sheath models and their impacts on sheath-driven dynamics at the plasma–liquid interface are presented. For a more comprehensive review of the role of the plasma sheath in plasma–liquid interactions, see [44].

5.3.1 Electrostatic double layer or Debye layer

As described in section 2.3.2, an electrostatic double layer or Debye layer forms at a water interface. The electrostatics of a plasma–liquid interface should thus be dictated by the formation of a Debye layer as well with an excess of ions or electrons on the plasma side of the interface balanced by solvated ions of opposite charge on the liquid side of the interface. Rumbach *et al* provided an analytic model for the electrostatic double layer formed at a plasma–liquid interface by combining a simple parabolic band model for the plasma sheath with the Gouy–Chapman theory (cf section 2.3.2) for the liquid [45]. The plasma sheath near the interface forms a counterpart in the liquid, composed of a compact layer of adsorbed ions and a diffuse layer governed by ionic strength and permittivity. The derivation of the analytic model provides lots of insights into the interplay between plasma and liquid and is thus provided in the following—all based on reference [45] where more details can be found. The general framework for their model is a plasma–liquid system with the plasma side being negatively biased towards the liquid. The liquid is an aqueous solution containing inert salt to provide conductivity. The sheath on the plasma side of the interface consists almost entirely of electrons diffusing and drifting from the quasineutral plasma bulk. The anode sheath region begins at $x = 0$ and ends at the solution interface $x = x_\mathrm{p}$. The electrostatic potential ϕ in the anode sheath is given by Poisson's equation:

$$\frac{\mathrm{d}^2\phi}{\mathrm{d}x^2} = \frac{qn_\mathrm{p}(x)}{\varepsilon_0} \tag{5.19}$$

Similar to equation (2.20) in section 2.3.2, a Taylor series about $x = 0$ is performed:

$$\frac{\mathrm{d}^2\phi}{\mathrm{d}x^2} = \frac{q}{\varepsilon_0}\left[n_{\mathrm{p}0} + n'_p(0)x + \frac{n''_\mathrm{p}(0)x^2}{2} + \cdots\right] \tag{5.20}$$

with $n_{\mathrm{p}0}$ the free electron density at the sheath boundary $x = 0$. With the assumptions that (i) electric field and potential go to zero in the plasma bulk at $x = 0$, (ii) the current density j in the anode sheath is spatially uniform and carried by electrons, (iii) $x < \lambda_\mathrm{D}$, (iv) small current density j, the electrostatic potential profile was estimated to be:

$$\phi(x) = \frac{qn_{\mathrm{p}0}}{2\varepsilon_0}x^2 \tag{5.21}$$

The total potential dropped across the sheath depends on the sheath length x_p:

$$\phi(x_\mathrm{p}) = V_\mathrm{s} = \frac{qn_{\mathrm{p}0}}{2\varepsilon_0}x_\mathrm{p}^2 \tag{5.22}$$

On the liquid side of the interface, there are positive and negative ions (e.g., Na^+, ClO_4^-) with a liquid bulk density n_l. The solution is charge neutral. Due to the electric field, positive ions are drawn towards the interface while negative ions are pushed away. A column of negative charges extends down from the interface due to solvated electrons and their recombination reaction (5.17), described above, forming OH^-. The potential on the liquid side can then be described by the Poisson–Boltzmann equation, cf equation (2.18):

$$\frac{d\phi^2}{dx^2} = \frac{zqn_l}{\varepsilon\varepsilon_0}\left[\exp\left(\frac{zq\phi(x)}{k_B T_l}\right) - \exp\left(\frac{-zq\phi(x)}{k_B T_l}\right)\right] + \frac{qn_{e,aq}}{\varepsilon\varepsilon_0} \approx \frac{\phi(x)}{\lambda_{D,l}^2} + \frac{qn_{e,aq}}{\varepsilon\varepsilon_0} \quad (5.23)$$

with T_l the temperature of the liquid, $\lambda_{D,l} = \sqrt{\frac{\varepsilon\varepsilon_0 k_B T_l}{2q^2 I_S}}$ the Debye length in the liquid, $I_S = z^2 n_l$ the ionic strength of an electrolyte, and $n_{e,aq}$ the interfacial concentration of solvated electrons which recombine to form OH^- (given as $n_{e,aq} \approx 0.5$ mM in [45]). The potential at the liquid side of the interface can then be obtained:

$$\phi(x) = -\left[V_l \exp\left(-\frac{x - x_p}{\lambda_{D,l}}\right) + \frac{k_B T_l}{2qz^2} \cdot \frac{n_{e,aq}}{n_l}\right] \quad (5.24)$$

where V_l is the voltage dropped across the liquid side of the interface. Inserting equation (5.24) into the Boltzmann equation yields the distribution of negative and positive ions in the solution:

$$n_\pm(x) = n_l \cdot \exp\left(\frac{\pm zq\phi(x)}{k_B T_l}\right) \approx n_l \cdot \left[1 \pm \frac{zqV_l}{k_B T_l} \cdot \exp\left(\frac{x - x_p}{\lambda_D}\right) \pm \frac{n_{e,aq}}{2z}\right] \quad (5.25)$$

It can be seen that positive ions are drawn towards the surface while negative ions are repelled, as expected. To couple solution chemistry and plasma physics, the boundary condition: $\varepsilon_0 \cdot E_p = \varepsilon\varepsilon_0 \cdot E_l$ is introduced to describe the displacement field at the plasma–liquid-interface. Inserting in equations (5.21) and (5.24) yields:

$$qn_{p0}x_p = \frac{\varepsilon\varepsilon_0 V_l}{\lambda_{D,l}} \quad (5.26)$$

The sheath thickness x_p and with that the sheath voltage V_s explicitly depend on the liquid bulk ion density n_l:

$$x_p = \frac{V_l}{n_{p0}}\sqrt{\frac{2\varepsilon\varepsilon_0 z^2 n_l}{k_B T_l}} \quad (5.27)$$

and

$$V_s = \frac{\varepsilon\varepsilon_0 V_l^2 qz^2 n_l}{\varepsilon_o n_{p0} k_B T_l} \quad (5.28)$$

The electric field at the interface increases with increasing ion concentration (ionic strength) in the liquid. The solution properties thus directly affect the electrostatic properties of the discharge. The analytical model was introduced for a DC glow discharge. Given the wide applicability of the Poisson–Boltzmann equation, regardless if pulsed discharges or AC discharges are used, the model should still hold true in general. Aqueous ions arrange themselves in a quasistatic equilibrium with the electric field. However, if the temporal changes in the electric field occur on much shorter timescales than the Maxwell relaxation time (e.g., rf or ns-pulsed discharges), the quasistatic assumption will fail. The strong electric field in the sheath also modifies the molecular and electronic structure of the interface. It can induce the reorientation of interfacial water dipoles and alter the electric double-layer structure by changing the ion distribution or dielectric alignment, which in turn can alter the permittivity and potential profiles in and near the interface [46, 47]. Significant accumulation of ionic and adsorbed species can affect the interfacial current and redox acitivity [45]. The adsorption of ions like OH^- can alter the local surface charge composition and density, affecting local redox activity.

5.3.2 Secondary electron emission from the liquid interface

While commonly discussed for solids (cf section 4.2), secondary electron emission from liquids is possible, especially under intense ion bombardment or high-field conditions. In a system where the liquid functions as a cathode, electrons are emitted from the liquid into the gas. These electrons may contribute to maintaining the plasma or influence local chemistry and boundary conditions. While the mechanisms of electron emission from solid surfaces are relatively well understood, the situation for liquids is more intricate. On the very short timescale of a plasma–surface interaction, the liquid surface can effectively behave like a solid (or ‘frozen’ in time), allowing some mechanisms known from solids to be extended to liquids. For instance, secondary electron emission may result from bombardment by ions, metastables (discussed below in section 5.4), or energetic neutrals, and can also be assisted by photoelectric or field-induced effects. Ion bombardment contributes both kinetic and potential energy, which can help liberate an electron.

Experimental and theoretical studies have shown that dielectric surfaces, including water ice, can have higher secondary electron emission yields compared to metals. While thermionic emission is generally unlikely for volatile liquids like water, strong local electric fields can support emission by inducing band bending and lowering the surface barrier.

However, several unique characteristics of liquids limit their electron emission. Unlike solids, charge transport in liquids occurs via drift, diffusion, and convection, which introduces transport limitations and may suppress emission. Furthermore, solvation dynamics and interfacial molecular organization can hinder the escape of electrons, making the emission event rare. Different underlying mechanisms for secondary electron emission were proposed, reviewed in [2]. Delgado *et al* provided one of the most detailed models to date, proposing that positive ions interacting with the liquid surface can ionize water molecules, generating conduction band electrons

[48]. The ions can transfer some of their kinetic energy to the surface water molecules, and if sufficient energy is provided ($E_{ion} \geqslant 11.16$ eV, [49]), ionize the liquid water molecules. As discussed in section 3.2.2, it is still under debate how much energy the ions have, given the highly collisional nature of atmospheric pressure plasmas. For low-energy ions the relative probabilities that an inelastic collision will result in excitation, ionization, or charge transfer are largely unknown, as collisional cross sections are not available. MD simulations have shown that the penetration depth of the ions, depending on their energy, ranges from an average of 7 Å for 10 eV ion energy to 15 Å for 100 eV ion energy, with the highest penetration depth of 27.5 Å [43]. This suggests that electrons would be created at or near the surface. These electrons may either escape into the plasma or rapidly localize into pre-solvated and eventually solvated states. The secondary electron emission coefficient γ describes then the probability of the combination of an ion ionizing a surface-near water molecule and a conduction band electron escaping. The authors note that this is rare event but not outside of the realm of possibility. Indeed, the upper limit for a secondary electron emission coefficient determined in their study is given as $\gamma \leqslant 10^{-5}$ [48]. The likelihood of this emission depends on the electron's position relative to the vacuum level and the energy provided by the ion. For instance, a conduction band electron lies approximately 1eV below the vacuum level, much more favorable for emission than a fully solvated electron, which requires over 3.5 eV to escape [48].

Alternative emission pathways have also been proposed [44]. For example, emission may arise from solvated electrons, negative ions near the interface, or even collective phenomena such as Auger-like transitions or exciton-plasmon coupling. More speculative theories involve quasi-particle excitations, such as plasmons formed at the interface, which could lead to electron emission if their energy exceeds the electron binding energy. These models highlight the possibility of discrete or collective emission events, potentially enhanced by transient field effects or surface structures like microdroplets or Taylor cones (more in section 5.5). Local field enhancement and vapor shielding may also alter the effective interaction site, shifting emission away from the main liquid surface toward clusters or droplets suspended above it. While various mechanisms have been proposed, including ion-induced ionization, solvated electron pathways, and quantum collective effects, conclusive evidence for dominant processes remains limited. Future experimental and computational work is needed to determine which pathways most significantly contribute to secondary electron emission at plasma–liquid interfaces.

5.3.3 Faradaic processes at the plasma–liquid-interface

At the plasma–liquid interface, Faradaic processes refer to electrochemical reactions that involve the transfer of charge, typically electrons, across the interface, resulting in a net chemical transformation. These processes are distinct from non-Faradaic phenomena such as double-layer charging or physical adsorption, as they involve redox reactions governed by interfacial potentials and plasma-driven fluxes. When charged species from the plasma interact with the liquid, they can induce electron transfer reactions that resemble classical electrochemistry but are often modulated by the unique characteristics of the plasma environment.

For example, plasma-supplied electrons can drive reduction reactions at the liquid surface. A common case is the reduction of metal cations dissolved in the liquid:

$$Ag^{+}_{aq} + e^{-} \rightarrow Ag_{s} \tag{5.29}$$

This reaction forms the basis of plasma-assisted electrochemical deposition, relevant in nanomaterial synthesis and plasma-enabled catalysis.

Conversely, when the plasma functions as an anode, positive ions can extract electrons from interfacial species, resulting in oxidation reactions. For instance, the oxidation of hydroxide ions can generate molecular oxygen and provide electrons:

$$2OH^{-} \rightarrow H_2O + \frac{1}{2}O_2 + 2e^{-} \tag{5.30}$$

A third representative Faradaic pathway involves the reduction of dissolved oxygen, commonly encountered in plasma medicine and plasma-activated liquids. Plasma electrons can reduce O_2 to hydrogen peroxide:

$$O_2 + 2H^{+} + 2e^{-} \rightarrow H_2O_2 \tag{5.31}$$

This reaction exemplifies how Faradaic electron transfer supports the generation of long-lived oxidants with biological relevance.

The occurrence and efficiency of these Faradaic processes are governed by multiple interdependent factors, including the local interfacial potential, plasma sheath structure, charged species fluxes, and the availability of redox-active species. Additionally, these reactions are coupled to transport dynamics such as diffusion, migration, and convection within the liquid phase and may be influenced by the transient and spatially varying electric fields near the interface. As such, Faradaic reactions at the plasma–liquid boundary form a critical link between plasma physics and interfacial electrochemistry.

The electrochemical potential determines the driving forces for redox reactions. In plasma–liquid systems, the electrochemical potential is strongly influenced by the sheath. The strong electric field in the sheath causes a sharp potential drop across a nanometer-to-micrometer scale and defines the interfacial voltage that drives electron or ion motion across the boundary. The flux of charged species alters the local charge density at the interface. For example, if the plasma supplies excess electrons, the interface becomes negatively charged, lowering the interfacial potential and favoring reduction reactions. Conversely, positive ion bombardment can locally increase the potential, favoring oxidation reactions. Pulsed or transient discharges induce rapid fluctuations in the potential, which can lead to periodic charging and discharging of the interface, potentially triggering non-steady-state Faradaic reactions.

These concepts find applications in plasma-driven solution electrochemistry, recently reviewed in [50].

5.4 Photon and metastable-induced interactions

At-a-glance
Atmospheric pressure plasmas supply VUV/UV photons and metastables whose effects are largely confined to a thin interfacial layer: photons are strongly absorbed (dose set by spectrum and liquid optics), while metastables are rapidly quenched near the surface. Photons drive water photoionization/photolysis to yield hydrated electrons and OH, and they photochemically activate solutes to produce secondary radicals. Metastables predominantly induce Penning-type ionization and dissociative energy transfer at the interface, with efficacy governed by interfacial composition (e.g., surfactants, structured water layers).

Plasma sources operating at atmospheric pressure emit photons across a wide spectral range, as described in section 3.2.3. The most chemically relevant for plasma–liquid interactions are VUV and UV photons, which carry sufficient energy to directly cleave chemical bonds, ionize water, and drive radical formation. These photons originate from excited species in the plasma, including atomic lines (e.g., 130 nm from O, 121.6 nm Lyman-α from H), excimer continua (e.g., 59–64 nm from first, and 64–100 nm from second He_2^* continuum, 106–115 nm and 115–165 nm from first and second Ar_2^* continua), and molecular bands (e.g., 98–102.5 from N_2), depending on the feed gas and discharge type. Photons in the VUV region are strongly absorbed by oxygen and water, with penetration depths ranging from hundreds of micrometers in the VUV, to millimeter and tens of millimeter scales in the UV-C and UV-B range [51]. This high absorption confines most primary photochemical events to a thin interfacial layer, making photon-induced chemistry highly surface-localized compared with plasma-generated reactive species that can diffuse deeper into the bulk. The photon dose within this layer depends on both the spectral distribution and the optical properties of the liquid, which are influenced by solutes such as dissolved organic matter or inorganic ions [52].

5.4.1 Primary photolysis and photoionization

If the high-energetic photons reach the liquid interface, they can initiate chemistry in water via photoionization [53]:

$$H_2O + h\nu \rightarrow H_2O + e^- \tag{5.32}$$

Direct ionization occurs for photon energies above 11 eV ($\lambda \lesssim 112$ nm), while lower photon energies require nuclear motion to favor electron ejection [54]. The electron rapidly becomes hydrated. These hydrated electrons are strong reductants (see above in section 5.3). Furthermore, high-energetic photons can photodissociate water via photolysis [55]:

$$H_2O + h\nu \rightarrow H_2O^* \rightarrow OH^\cdot + H^\cdot \tag{5.33}$$

Hydroxyl radicals ($OH^{\cdot}$) are potent oxidants, initiating chain reactions with organics and inorganics. However, many of the primary species generated undergo geminate recombination[1] within femtoseconds to nanoseconds [54], reducing effective yields. The interplay between recombination and scavenging determines net radical and electron concentrations.

5.4.2 Secondary solute-driven reactions

Photons may also be absorbed by solvated oxygen and other solutes rather than water, producing secondary radicals or excited states. For example (more reactions can be found in [57]):

$$O_2 + h\nu \rightarrow 2O^{\cdot} \tag{5.34}$$

If the solution contains other components, such as naturally occurring organic compounds, sodium chloride in ringer-lactate solutions, or either pre-existing or plasma-produced nitrites and nitrates, the molecules may dissociate upon absorption of UV radiation, leading to secondary reactions in the aqueous environment. Note, that the impact of VUV radiation is then limited to surfactants, as VUV is not able to penetrate the bulk liquid. Halide ions (X^-), such as chloride, bromide, or iodide ions, absorb UV radiation <260 nm, where the halide ion forms a halogen atom–electron solvent cage complex:

$$X^- + h\nu \leftrightarrow (X^{\cdot}, e^-) \tag{5.35}$$

This cage complex can either recombine: $(X^{\cdot}, e^-) \rightarrow X^-$ or undergo a chemical reaction with the solvent water:

$$(X^{\cdot}, e^-) + H_2O \rightarrow X + H + OH^- \tag{5.36}$$

to form hydrogen and halogen atoms as well as the OH ion [58, 59].

Nitrate and nitrite photodissociate upon UV absorption to produce reactive nitrogen oxides and atomic oxygen:

$$NO_2^- + h\nu \rightarrow NO^{\cdot} + O^{\cdot -} \tag{5.37}$$

$$NO_3^- + h\nu \rightarrow NO_2^{\cdot} + O^{\cdot -} \tag{5.38}$$

$$NO_3^- + h\nu \rightarrow NO_2^- + O^{\cdot} \tag{5.39}$$

These dissociation products will immediately react further in the aqueous environment in secondary reactions.

Hydrogen peroxide can be dissociated with UV-C radiation to produce OH, a well-known advanced oxidation process [1]:

$$H_2O_2 + h\nu \rightarrow 2\ OH^{\cdot} \tag{5.40}$$

[1] Geminate recombination refers to the reaction of two transient species produced from a common precursor in solution with each other before a separation by diffusion could occur [56].

Lastly, photons can just transfer their energy to the solution which can result in local energy deposition and localized heating.

5.4.3 Metastable-induced interactions

Similarly, metastables can deposit energy into the solution. The dense environment near the liquid surface strongly quenches metastables through collisions with vapor-phase water molecules and dissolved gases. The probability of direct metastable penetration into the liquid bulk is extremely low due to rapid deactivation; instead, most chemical influence is exerted within the first few monolayers or via secondary species produced at the interface. The interaction of metastables with liquid surfaces is less extensively documented, but Harada *et al* [60] briefly note that the deexcitation mechanisms are believed to be similar to those on solid surfaces (cf section 4.3), particularly for organic liquids. Penning ionization is the dominant deexcitation process, especially if the liquid's molecules behave like loosely bound organic films. The interaction of metastables with liquid surfaces will depend on the surface composition (e.g., presence of surfactants, structured water layers), the orientation and density of molecular orbitals at the liquid interface, and the availability of loosely bound electrons or low-energy ionization pathways.

In practice, metastables approaching a liquid interface likely ionize or de-excite through similar Penning-type interactions, but with less predictable orientation and more dynamic surface behavior than on rigid solids. Metastables can efficiently produce ions and electrons in the interfacial gas phase, which can initiate secondary reactions in the adjacent liquid. Dissociative energy transfer may produce reactive species from interfacial molecules, such as OH from water. Furthermore, the excitation of surface or near-surface molecules can lower activation barriers for subsequent chemical reactions, including those with radicals or photons.

5.5 Plasma-induced fluid motion and interfacial dynamics

At-a-glance

Beyond chemistry, plasmas exert mechanical forces on liquids, driving flows, oscillations, and instabilities that reshape the interface and enhance transport.

- Driving mechanisms: Ion momentum, electron pressure, Maxwell stresses, Marangoni flows, and gas shear each contribute, depending on plasma type and liquid properties.
- Instabilities and deformations: Electric fields and surface tension interplay can create cone-like protrusions, droplet ejection, or oscillations that couple fluid motion with species transfer.

The effect of plasma on liquid surfaces is not just driven by species transport and chemical reactions, but also by mechanical forces and stress gradients that drive liquid flows and shape the interface. These can lead to steady flows, oscillations, and instabilities that reshape the interface. The physical motion of the interface or the bulk liquid can then influence mass transfer, heat transport, and species delivery. A variety of physical concepts have been suggested as driving forces of plasma-induced fluid motion, mainly because the dominating driving force depends significantly on the plasma source and the liquid. In a direct discharge with an electric field normal to the liquid surface, the capillary pressure difference sustained between the two fluids plasma and water does now not only depend on the surface tension (cf section 2.3.1, Young–Laplace equation (2.9)). Charged particles and the electric field add pressure to the liquid surface via ion momentum transfer, electric Maxwell stress, and electron thermal pressure. Localized energy deposition and heating can cause temperature gradients, while chemical reactions can create concentration gradients, both impacting surface tension and liquid flow. In the following, an overview of the different mechanisms is given in table 5.4, followed by the description of the driving mechanisms of plasma-induced flows in section 5.5.1 and a discussion on deformation, oscillations and instabilities in section 5.5.2.

5.5.1 Driving mechanisms of plasma-induced flows

5.5.1.1 Marangoni flows

Gradients in surface tension due to temperature (∇T) or concentration (∇c) gradients lead to Marangoni flow (cf section 2.3.1). Heating of the liquid due to ion impact (cf section 5.3) or metastable and UV/VUV photons (cf section 5.4), or Ohmic heating can cause a temperature gradient along the liquid surface resulting in a gradient in surface tension that causes a hydrodynamic drag, leading to a flow of the liquid—the thermocapillary Marangoni flow. Similarly, plasma species adsorbing and desorbing at the interface can change the concentration locally [61]; this gradient in concentration will cause a gradient in surface tension, leading to the solutal or soluto-capillary Marangoni flow. Research by Kawasaki *et al* has shown a change in surface tension after plasma treatment of a surfactant solution and that the change in surface tension was strongly related to the direction and intensity of the liquid flow [62]. For low surfactant concentrations, the flow along the axis of the gas flow increased. After a certain initial surfactant concentration, the flow pattern reversed and flow weakened [62]. Similarly, Thagard *et al* observed a flow reversal (from away from the discharge location to towards the discharge location) for solutions with strong surfactants such as perfluorooctanoic acid [63]. As the surfactant molecules degrade over time, the surface tension gradient is reduced and the driving force for Marangoni flows diminishes. Thus, the contributing factors for Marangoni flow can be transient during treatment, with changing temperature and concentration gradients that lead to dynamic surface gradients and changing flow patterns.

5.5.1.2 Electrohydrodynamic flows

Liquid surfaces experience a net force inward due to surface tension, as described in section 2.3.1, leading to the Young–Laplace capillary pressure. When a liquid

Table 5.4. Overview of key driving mechanisms of plasma-induced liquid flows, their physical origins, dominant parameters, and example observations.

Mechanism	Physical origin	Key parameters	Typical timescale	Example observations
Ion momentum transfer (ram pressure)	Sheath-accelerated ions transfer momentum at stagnation point on liquid surface	Ion density n_i, ion mass m_i, ion velocity u_i, sheath potential ϕ_s	μs–ms	Steady circulation beneath impact zone; protrusions or dimples
Electron thermal/Maxwell stress	Electron pressure and electric field gradients exert normal stress on surface	T_e, E, permittivity ε, conductivity σ	μs–ms	Localized bulging, suppression or triggering of surface instabilities
EHD deformation	Coulomb forces and Maxwell stress deform liquid interface	E, ε, σ, viscosity μ	μs–ms	Lifting/dimpling of surface; cone-like protrusions
Marangoni (thermo-/soluto-capillary)	Surface tension gradients due to ∇T or ∇c along interface	$\partial\gamma/\partial T$, $\partial\gamma/\partial c$, temperature gradient, concentration gradient	ms–s	Flow reversal in surfactant solutions; outward/inward surface flows
Gas flow convection	Tangential shear from deflected gas jet flow	Gas velocity u_j, gas density ρ_g, density ratio $\rho_\mathrm{l}/\rho_\mathrm{g}$	ms–s	Dimple formation, splashing, penetration events

surface is exposed to an electric field and charged species, additional pressure is acting on the liquid:

$$p_s = m_i n_s (\underline{u}_s \cdot \underline{\hat{n}})^2 + n_0 k_B T_{e,s} \exp\left(\frac{e\phi}{k_B T_{e,s}}\right) - \frac{\varepsilon_{eff} E_s^2}{2} - \gamma(-\nabla \cdot \underline{\hat{n}}) \quad (5.41)$$

with the subscript s denoting values at the surface, and $\underline{\hat{n}}$ the surface unit normal pointing towards the plasma. The first term on the right-hand side denotes the ion momentum transfer, or to phrase it in fluid dynamic terms: ion ram pressure. In some publications, the momentum transfer of charged species has been suggested as the main driving force for plasma-induced liquid flow [64]. The second term describes the electron thermal pressure, the third term is due to electric Maxwell stress, and the fourth term describes the Young–Laplace pressure with surface tension γ. The electric Maxwell stress contains the frequency-dependent effective permittivity which includes the conductivity of the medium via the Maxwell relaxation time $\tau_M = \varepsilon/\sigma$ (cf equation (5.16)):

$$\varepsilon_{eff} = \varepsilon \frac{1}{1 + \omega^2 \tau_M^2} \quad (5.42)$$

The permittivity can be approximated as $\varepsilon_{eff} \approx \varepsilon$ if $\omega^2 \tau_M^2 \ll 1$. For example, if the liquid is de-ionized water with a conductivity of 0.2 μS cm^{-1} and $\varepsilon_r = 80$: $\tau_M = \varepsilon_0 \varepsilon_r / \sigma \approx 35$ μs. The approximation is then valid for frequencies $\omega \ll 30$ kHz.

Holgate *et al* [65] used the pressure balance in equation (5.41) to couple numerical simulations and an EHD code to solve the dynamics of an incompressible liquid under the pressure gradients generated by surface tension, electric fields, ions, and electrons at the liquid surface. It was found that the liquid circulates steadily under the pressure gradients generated by the plasma according to equation (5.41), in accordance with experimental observations [66–68]. By fully accounting for the strong electric fields and particle flows in the sheath region that forms at the interface, sheath effects were found to be crucial for the shape and stability of the liquid surface. Electric fields in the sheath can deform the surface into a cone-like protrusion (see next section 5.5.2), up to the ejection of a burst of droplets from its apex. Ions, accelerated in the sheath, can provide a countering force on the liquid surface that can suppress EHD instabilities [65]. (Note: the ion wind effect, sometimes described separately, is often treated as part of ion momentum transfer in the sheath, but in some contexts may also be modeled as shear-driven flow.)

In their work, Holgate *et al* use the dimensionless plasma bond number Bo_p that compares the relative importance of plasma pressure to capillary pressure of the liquid [65]:

$$Bo_p = \frac{n_0 k_B T_e \lambda_D}{\gamma} \quad (5.43)$$

$Bo_p \ll 1$ means surface tension dominates over plasma pressure, $Bo_p \gg 1$ means plasma pressure can strongly deform the surface.

A different non-dimensional number that could be adapted from fluid dynamics is the capillary number which, in classic fluid dynamics, gives the ratio of viscous forces to surface tension forces in a fluid system. A plasma capillary number Ca_{p} could take the form:

$$Ca_{\mathrm{p}} = \frac{\sum p_{\mathrm{plasma}}}{\gamma/R} = \frac{\varepsilon_{\mathrm{eff}} E^2 R}{2\gamma} + \frac{n_{\mathrm{e}} k_{\mathrm{B}} T_{\mathrm{e}} R}{\gamma} + \frac{m_{\mathrm{i}} n_{\mathrm{i}} u_{\mathrm{i}}^2 R}{\gamma} \tag{5.44}$$

The first term on the right side describes the Maxwell stress capillary number, the second the electron pressure capillary number, and the third term the ion capillary number. The plasma capillary number can be used to describe bulges, cones, and dimples when the expression γ/R is used for quasistatic curvature. For waves and perturbations, R can be replaced by $1/k$.

5.5.2 Interfacial deformations, oscillations, and instabilities

5.5.2.1 Electrohydrodynamics deformation

Liquid surfaces are deformable and EHD forces can alter the shape during plasma–liquid interactions. Weak and stable deformations can be observed when the Maxwell electric stress is not exceeding the capillary pressure at a curvature scale $R \sim 1/k$ (neglecting gravity and viscosity):

$$\frac{\varepsilon_{\mathrm{eff}} E^2}{2} \leqslant \gamma k \Rightarrow E \leqslant \sqrt{\frac{2\gamma k}{\varepsilon_{\mathrm{eff}}}} \tag{5.45}$$

When this holds true, small perturbations are restored and the system is stable. It was observed in a pin-to-liquid electrode system, for example, that a small Coulomb force was induced for electric fields below the Corona breakdown voltage, lifting the water up [69]. Above the breakdown voltage a larger repulsive force was created due to ionic wind that depressed the water several hundred micrometers in the center [69]. In another metal–liquid electrode system, the elevation height of the water before electrical breakdown was shown to increase with applied voltage [70]. The elevation of the water results in a flow in the liquid phase against gravity [71]. The deformation of the interface also depends on the liquid used. For de-ionized water, the liquid behaves like a lossy dielectric and Maxwell electric stresses become more important [71]. For a slightly more conductive medium, such as tap water, the EHD-induced gas flow or ion wind becomes more important [71]. For the de-ionized water case, larger deformations of the interface were observed [71]. Interestingly, in a plasma jet system with gas flow (more about gas flow dynamics and instabilities in section 5.6), the EHD forces were shown to increase the depth of the dimple created by the gas flow, forming a longer cavity [72]. Furthermore, the EHD forces stabilized liquid instabilities.

5.5.2.2 (Non-)capillary waves and Rayleigh–Taylor instability

Capillary waves are small-amplitude oscillations at the liquid–gas interface, in which surface tension acts as the principal restoring force, especially for short-wavelength disturbances. In nature, wind can stir up small ripples and on the open ocean, the much larger ocean surface waves may be a result of coalescing smaller wind-caused capillary waves.

When Maxwell stress and capillary pressure are near balance, the field modifies the dispersion (capillary term is partly neutralized) so waves and oscillations are easy to excite. Then, similar to equation (5.45),

$$E \sim \sqrt{\frac{2\gamma k}{\varepsilon_{\text{eff}}}} \tag{5.46}$$

Rayleigh–Taylor instabilities can arise when two fluids of different density interface each other, where the denser fluid experiences acceleration into the lighter fluid in the presence of interfacial perturbations. Maxwell normal stress and ion momentum transfer can act like an effective 'acceleration' of the heavy liquid toward the lighter gas, so it could be defined as an EHD Rayleigh–Taylor-like instability.

A conductivity-dependent analysis of an EHD Rayleigh–Taylor-like instability predicted that the growth rate of the instability increases with decreasing conductivity of the liquid [73]. An EHD Rayleigh–Taylor-like instability pathway can emerge when the sheath's normal electric stress exceeds the capillary and (if relevant) hydrostatic restoring forces over a band of wavenumbers. For small-amplitude perturbations of wavenumber k on a flat interface between a viscous liquid (index l) and a gas/plasma (index g), a growth-rate model in the leaky dielectric/Stokes limit can be written as:

$$\sigma(\omega, k) \approx \frac{k}{\mu_{\text{l}} + \mu_{\text{g}}}[\Delta p_{\text{E}}(\omega) - \gamma k^2 - \Delta \rho g] \tag{5.47}$$

with $\sigma(k)$ the temporal growth rate (instability if $\sigma(k) > 0$), μ_i the viscosities, Δp_{E} the electric Maxwell stress, and $\Delta\rho = \rho_{\text{l}} - \rho_{\text{g}}$.

The field distortion at the plasma–liquid interface depends on the conductivity and k. Lower conductivities mean higher Maxwell relaxation times, which increases the effective permittivity and results in a larger destabilizing Δp_{E}, consistent with results reported in [73]. The growth intensifies with increasing normal electric field and decreasing conductivity, and is suppressed by higher surface tension and viscosity.

Recently, a new class of surface waves have been proposed that arise from resonant coupling of discrete plasma pattern modes and a continuum of interfacial surface wave modes [74] highlighting a close coupling between plasma-induced electric patterns and surface wave modes, leading to nonstandard, non-dispersive wave behavior that differs from purely thermal capillary fluctuations. These non-capillary waves are entirely dictated by EHD interactions rather than surface tension. The surface tension effects became neutralized by the intense electric field across the plasma–liquid interface. A new dispersion relation was derived in [74]:

$$\omega^2 = \frac{qn_0\lambda_D E_0}{\rho}k^2 \tag{5.48}$$

The experimental results with a liquid anode glow discharge system align well with the theoretical predictions and even the pattern formation could be predicted [74]. The size of the patterns is dictated by plasma processes while the transient behavior of the patterns is limited by the advection of solvated salt ions via viscous surface waves. The phase velocity of the noncapillary waves was determined to be 0.08 m s^{-1}, an order of magnitude lower than the traditional capillary wave theory would predict. While the newly proposed theory represents the experimental data very well, the authors also note that there are many other EHD effects that should be considered, such as electrostatic shear stress and electromigration of solution ions [74]. Understanding these wave phenomena may offer new insights into interface deformation, reactive species transport, and pattern formation in plasma–liquid systems. A computational investigation of a structured liquid interface (assumed to be produced by acoustic shaping) with a standing wave pattern in contact with a surface ionization wave has shown electric field enhancements at the peaks of the waves, producing local increases in reactive oxygen and nitrogen species (RONS) and incident plasma fluxes [75]. This would allow for position-dependent material synthesis, for example, in plasma-driven solution electrochemistry [75].

5.5.2.3 Electrohydrodynamic cone and jet formation (Taylor cones)

Once the local field exceeds the cone onset scaling,

$$E \gtrsim \sqrt{\frac{2\gamma}{\varepsilon_{\text{eff}}R}} \tag{5.49}$$

and/or when the electric pressure dominates capillarity over a band of wavenumbers,

$$E \gtrsim \sqrt{\frac{2\gamma k}{\varepsilon_{\text{eff}}}} \tag{5.50}$$

sustained cone-jetting or strong cavity growth can be observed.

The EHD cone, commonly referred to as the Taylor cone[2], arises when a conducting liquid is exposed to a sufficiently strong electric field. Under these conditions, the balance between surface tension, which favors a smooth interface, and electrostatic stresses, which pull the liquid outward, leads to the formation of a conical meniscus that develops into cone-jetting when the electric field is high enough.

Taylor's seminal analysis (1964) went further, showing that the equilibrium semi-angle of a perfectly formed cone approaches 49.3°, a value derived from the full balance of electrostatic and capillary forces [79]. Earlier, Rayleigh had described the

[2] Some researchers have (unsuccessfully) attempted to rename the dynamic electrified protrusions as Gilbert–Gray cone-jets [76]. Gilbert was the first to report the attraction of water to a charged piece of amber, followed by the emission of a fine jet, whereas Gray reported similar jet formations in water and liquid mercury [77, 78].

instability and breakup of charged droplets, laying the groundwork for understanding droplet emission from cones. Later refinements by Melcher and others included viscous and dielectric effects, broadening applicability to real liquids beyond the ideal conductor case [80].

A strong electric field reorders the dipolar orientation of water molecules until a protrusion develops into a Taylor cone-like structure [81]. For pure water, a columnar structure parallel to the field develops, with the dipole moments of the water molecules aligned along the applied electric field. In electrolyte solutions, the ion–water interaction interferes with the water–electric field interaction and the competition between electric field strength and ionic concentration drives the stability of the observed structures. Depending on the conditions, the Taylor cone-like structure can further grow into a vertical water bridge. Depending on the plasma configuration and liquid properties, the surface may undergo small deformations, evolve into a stable cone, or emit microjets and droplets. Time-resolved imaging has revealed both transient bulging and sustained cone formation under continuous discharge conditions [70]. Streamer formation can take place between the water tip and the electrode, until the water bridge connects to the upper electrode and shortcuts the circuit. The strong local electric field at the tip of the water cone can also lead to the emission of charged droplets [82] which then become part of the discharge [1, 2].

The instabilities have a cone-line structure. However, cones are not always formed as it also depends on the hydrodynamic and charge relaxation times. If $\tau_{\mathrm{M}} = \sqrt{\varepsilon/\sigma} \ll \tau_{\mathrm{H}} = \mu L/\gamma$, the interface behaves like a leaky conductor and sustained cone-jetting is plausible above the critical electric field.

5.5.2.4 Kelvin–Helmholtz instability

In addition to oscillations driven by electric fields, velocity shear at the interface can trigger instabilities such as the Kelvin–Helmholtz instability [83]. The Kelvin–Helmholtz instability can be observed throughout the Universe, forming cloud formations on Earth as well as on Saturn and Jupiter, among others [83]. The Kelvin–Helmholtz instability is well studied in magnetohydrodynamics, but less explored in atmospheric pressure plasma–liquid interactions. A stability analysis by Lai *et al* for a 2D bubble plasma setup in water has found that the velocity shear observed experimentally is due to surface tension and fluid flow initiated by a streamer head, but that the velocity shear regions are stable [66]. The instability growth rate scales with wavenumber k. For short wavenumbers, the modes satisfying $k\Delta < 2$ are unstable, with Δ the length scale of shear and $k = 2\pi/\lambda$, λ being the wavelength of perturbation observed at the shear interface. With experimental observations of $\Delta \approx 1.5$ mm and $\lambda \approx 2.7$ mm, $k\Delta \approx 3.5$, and thus stable. However, the stable boundary can still create vortices that can increase mixing and convection flow in fluids [66]. In a computational study with a plasma jet setup, it was found that pulsed heating launches acoustic waves that seed shear instabilities, causing more mixing of the gas downstream with ambient air [84]. However, the study did not address a liquid interface.

5.6 Differences between direct and indirect plasma sources

At-a-glance
Direct plasmas deliver charged particles and photons to the interface, while indirect sources mainly supply neutral species; this distinction fundamentally alters interfacial chemistry and transport pathways.

As outlined in sections 3.1.3 and 4.4 there are certain differences between direct and indirect plasma sources. Similarly to the solid interface, direct plasma sources such as volume DBDs and ionization wave jets will couple electrically to the liquid, providing a path for feedback and interactions. The electrical characteristics of a liquid, such as its conductivity and permittivity, define the boundary conditions that govern electric field distributions at the interface. Polarizable liquids will orient their molecules along the electric field lines. The charge relaxation time of the liquid τ_M determines surface charge accumulation, giving rise to electric surface stresses responsible for large deformations of the interface when the liquid behaves more like a dielectric [71]. Unlike solids, liquid surfaces can deform and flow in the liquid can be induced. In the presence of dissolved salts in the water, ion concentrations may vary causing density, conductivity, and permittivity gradients [52]. When the liquid is behaving more like a dielectric, approaching ionization wave jets will continue on the liquid surface as surface ionization waves. In this scenario, the contact area of the ionization wave with the liquid increases dramatically, and thus, exposure to ions, electrons, short-lived reactive species, metastables, and VUV/UV photons.

Certain jet types can be operated in a touching (direct) and non-touching (indirect) mode, depending on the distance between the jet and the target. These discharge types allow for a direct comparison of direct and indirect discharges and have been used to explore interactions of plasma jets with liquid water [85]. Before going into the comparison of touching and non-touching jets interacting with liquids, a brief introduction of just gas flow interactions (without plasma) with liquids is necessary.

5.6.1 Gas flow dynamics and indirect jets

A gas jet impinging on a liquid surface can deform the liquid surface and lead to complex surface dynamics. Depending on the gas flow and distance, dimpling, splashing, and penetration can occur where splashing and penetration are accompanied by oscillations and rotations [86]. At low gas flow rates, the gas flow will just induce a small depression of the surface, a dimple, of the liquid. With increasing gas flow, splashing occurs: the dimple will penetrate further into the liquid and generate surface waves, while oscillating between minimum and maximum penetration depth. As the gas flow is further increased, penetration occurs: the dimple penetrates deeper into the liquid while oscillating and rotating or bending. A great overview with illustrations of each mode is provided in [86]. With a gas flow alone, the depth of the cavity formed by the gas flow depends on the jet velocity. A critical velocity u_{cr} for the onset of splashing is provided in [86] as:

$$u_{\mathrm{cr}} = \left(\frac{48g\gamma(\rho_{\mathrm{l}} - \rho_{\mathrm{g}})}{\rho_{\mathrm{g}}^{2}} \right)^{0.25} \tag{5.51}$$

Together with the criterion that the blowing number $N_{\mathrm{B}} > 1$ for the onset of splashing based on the Kelvin–Helmholtz instability criterion, with N_{B} defined as:

$$N_{\mathrm{B}} = \frac{\rho_{\mathrm{g}}(\eta u_{\mathrm{j}})^{2}}{2\sqrt{\sigma_{\mathrm{l}} g \rho_{\mathrm{l}}}} \tag{5.52}$$

with u_{j} the gas jet velocity and $\eta = 0.44721$ a constant [86]. When the gas jet is impinging on the liquid surface, the gas flow gets tangentially deflected at the stagnation point and flows outwards along the air–liquid interface, depicted in figure 5.2 on the left. This flow pattern induces an upward flow in the liquid owing to friction. The liquid is accelerated towards the sides, leading to reversed vortices. Gas-flow-induced liquid flow (convection) is applicable for both direct and indirect

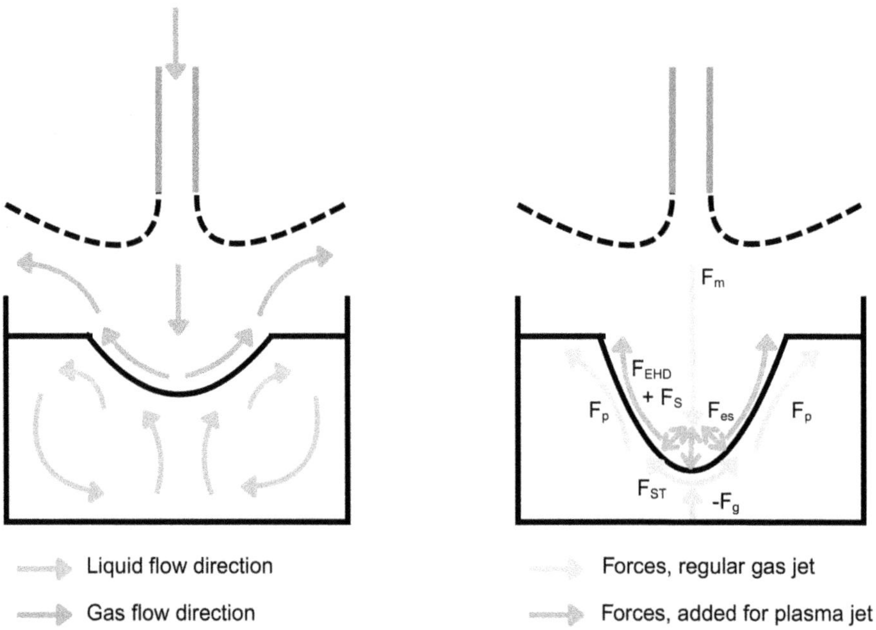

Figure 5.2. Left: Gas flow and liquid flow patterns for a typical direct jet-water setup. The incoming gas flow (orange) approaches the liquid interface, causing a dimple. At the stagnation point, the gas flow gets redirected and flows outwards along the air–liquid interface. This flow pattern induces an upward flow in the liquid. Right: Forces acting on the liquid surface. Forces in a pure gas system are indicated by yellow arrows, the green arrows represent additional forces in a plasma–liquid system. Redrawn based on [72], with permission from Springer Nature.

jet sources and is an important factor as it plays a dominant role in the transport of species for efficient mixing. A recent study has investigated the transport of species from the gas phase to the liquid and within the liquid for different gas flow rates with a COST jet-similar device [28]. At a gas flow rate of 2 slm, a dimple was observed and velocities in the liquid reached 3 mm s^{-1}. For gas flow rates of 0.5 slm, no dimple, vortex pattern, or liquid motion was detectable and the transport occurred to a larger extent diffusion-based. At 1 slm, no dimple was observed and the liquid motion dropped to 1 mm s^{-1}, showing a mix of diffusion and convection [28]. The Péclet number is a dimensionless quantity defined as the ratio of convection and diffusion and helpful for computational studies and to estimate in what regime the jet is operating [26]:

$$Pe = u_* L_* / D \tag{5.53}$$

with u_* the characteristic velocity normal to the liquid surface, and L_* the characteristic width of the layer, and D the diffusion coefficient. $Pe \ll 1$ means the transport is diffusion-driven, $Pe \gg 1$ means the transport is convection-dominated. Semenov *et al* calculated the Péclet number in a computational transport study for the kINPen with typical values $u_* \lesssim 1\,\mathrm{m\,s^{-1}}$ and $L_* \sim 1\,\mathrm{mm}$ as $Pe \lesssim 40$ [26].

5.6.2 Gas flow dynamics and direct plasma jets

For gas jets impinging on a liquid surface it has been shown that the force balance is more accurate than a pressure balance at the interface [87], so that forces are used in this section instead of pressure like in the previous sections. When a touching jet or direct plasma jet source is in contact with a liquid surface, the interface is exposed to forces in addition to the momentum transfer from the gas jet F_m: electrostatic F_{es} and EHD F_{EHD} forces, so that the total force F_{ex} exerted on the liquid surface becomes [72]:

$$F_{ex} = F_m + F_{es} + F_{EHD} \tag{5.54}$$

Indeed, it has been observed experimentally that the cavity expands further along the jet axis when a plasma jet is impinging on the liquid, e.g., in [72, 85]. In addition to the electrostatic and EHD forces, an increased gas temperature can lead to an increased gas velocity which in turn increases the momentum transfer to the liquid surface. Kondeti and Bruggeman also found that the dimple oscillation frequency correlates with filament attachment dynamics when the plasma jet is at a short (touching) distance. For longer distances (non-touching), the oscillation amplitude and frequency depends on the container size of the liquid [85]. The dimple oscillations induce a liquid flow that can enhance mixing of the liquid and transport of species. In their work, the dimple-induced velocity in the liquid was estimated to be between 10 and 40 mm s^{-1} [85].

The additional forces acting on the liquid surface can also assist in stabilizing the interface, as reported in [72] and depicted in figure 5.2 on the right. Forces in a pure gas system are indicated by yellow arrows, the green arrows represent additional forces in a plasma–liquid system. In a pure gas system, the force balance at the interface is given by $F_m = k(F_g + F_{ST})$, with the coefficient k being an empirical shape factor [72]. The momentum transfer from the gas jet is balanced by the sum of the forces due to gravity (buoyancy) F_g and surface tension F_{ST}. The forces F_p and F_s are representing the rotational flow pressure and tangential shear stress, respectively. The plasma jet causes an increased force F_m due to gas heating in the discharge, and an EHD force F_{EHD} as well as an electrostatic force F_{es}. Then, the force balance at the interface is given as: $F_m + F_{EHD} + F_{es} = k(F_g + F_{ST})$. The interaction of the plasma with the liquid inside the cavity allows for bidirectional flow and an electric field parallel to the water surface, causing cavity expansion and stabilization [72]. Table 5.5 gives an overview of the different aspects and their differences for direct and indirect discharge concepts.

Table 5.5. Direct versus indirect plasma plasma sources interacting with liquids. Direct sources establish electrical contact and a sheath at the interface; indirect sources primarily act through gas flow and long-lived species.

Aspect	Direct (touching)	Indirect (non-touching)
Electrical coupling	Strong electrical coupling; surface charges and displacement currents close through the liquid.	Little/no electrical coupling; target floats electrically with only weak capacitive effects.
Sheath at interface	Well-defined plasma sheath at the liquid; strong normal E at the surface.	No sheath on the liquid; fields largely confined.
Dominant additional forces	Electrostatic (Maxwell) normal stress and EHD body forces in the near-surface liquid.	Primarily hydrodynamic shear from the gas jet; EHD effects negligible at the interface.
Interface deformation	Larger, deeper dimples/cavities; cone-like protrusions possible; suppression or triggering of instabilities by E-field and ion momentum.	Dimple/splash/penetration set mainly by jet momentum and fluid properties; deformation weakens with standoff distance.
Species delivery	High flux of short-lived species (e^-, ions, O, etc) metastables, and UV/VUV at the surface; strong local chemistry.	Dominated by long-lived species (O_3, NO_x, H_2O_2 vapor) transported by convection/ diffusion.
Thermal effects	Local gas heating can increase u_j and deepen cavity; Joule/ion heating at interface.	Lower interfacial heating; gas temperature closer to feed/ ambient conditions.
Marangoni/solutal effects	Strong: plasma alters ∇T and ∇c directly at the surface; possible flow reversals with surfactants.	Present but typically weaker; driven by composition/ temperature gradients from the impinging effluent.
Stability/oscillations	Dimple oscillations often correlate with filament attachment; E-field can stabilize cavity edges.	Oscillation amplitude/frequency set by flow and container geometry; less sensitive to discharge dynamics.

Typical dimensionless cues	Larger effective Bo_p (plasma pressure versus capillarity); small $\tau_M = \varepsilon/\sigma$ favors leaky-dielectric EHD.	Péclet number $Pe = u_* L_* / D$ governs transport; Kelvin–Helmholtz criteria relevant for splashing onset.
Sensitivity to liquid properties	Strong dependence on σ, ε, γ, and viscosity μ; conductivity sets charge relaxation and field penetration.	Strong dependence on γ, ρ_l/ρ_g, μ; weaker dependence on σ and ε.

Chapter 5—summary

- Plasma–liquid systems are highly diverse (gas-contact, liquid discharges, multiphase), and their dynamics depend strongly on discharge geometry, liquid properties, and timescales of charge relaxation.
- Neutral species (radicals, RONS) dominate plasma–liquid chemistry, with interfacial adsorption, solubility, and residence times determining whether reactions occur at the surface or in the bulk.
- Transport across the interface is not trivial; multiple models exist (Henry's law, kinetic, diffusive, film theory, reaction–diffusion), with accuracy depending on solubility, timescales, and plasma-driven non-equilibrium effects.
- Charged species influence interfacial electrochemistry through double-layer formation, solvated electrons, secondary electron emission, and Faradaic processes, tightly coupling plasma physics with electrochemistry.
- Photons (UV/VUV) and metastables act mainly at or near the interface, driving ionization, photolysis, and surface-localized chemistry with very short penetration depths.
- Plasma-induced mechanical effects (EHD flows, Marangoni effects, interfacial instabilities) reshape the liquid surface and strongly impact mass transfer, mixing, and chemical outcomes.

Reading recommendations

- More about plasma–liquid systems, plasma physics of liquids and with liquids: [1–3].
- Chemical reaction networks for plasmas inside liquids: [10, 11].
- Plasma–aerosol interactions: [14, 18].
- Plasma sheath in plasma–liquid interactions: [44] and plasma sheath formed around droplets: [88].
- Plasma-driven solution electrochemistry: [50].

References

[1] Bruggeman P J, Kushner M J, Locke B R, Gardeniers J G E, Graham W G, Graves D B, Hofman-Caris R C H M, Maric D, Reid J P and Ceriani E 2016 Plasma–liquid interactions: a review and roadmap *Plasma sources science and technology* **25** 053002

[2] Vanraes P and Bogaerts A 2018 Plasma physics of liquids–a focused review *Appl. Phys. Rev.* **5** 031103

[3] Bruggeman P and Leys C 2009 Non-thermal plasmas in and in contact with liquids *J. Phys. D: Appl. Phys.* **42** 053001

[4] Sponsel N L, Gershman , Quesada M J H, Mast J T and Stapelmann K 2022 Electric discharge initiation in water with gas bubbles: A time scale approach *J. Vac. Sci. Technol.* A **40** 063002

[5] Dobrynin D, Seepersad Y, Pekker M, Shneider M, Friedman G and Fridman A 2013 Non-equilibrium nanosecond-pulsed plasma generation in the liquid phase (water, pdms) without bubbles: fast imaging, spectroscopy and leader-type model *J. Phys. D: Appl. Phys.* **46** 105201

[6] Shneider M N, Pekker M and Fridman A 2012 Theoretical study of the initial stage of sub-nanosecond pulsed breakdown in liquid dielectrics *IEEE Trans. Dielectr. Electr. Insul.* **19** 1579–82

[7] Shneider M N and Pekker M 2013 Cavitation in dielectric fluid in inhomogeneous pulsed electric field *J. Appl. Phys.* **114** 214906

[8] Shneider M N and Pekker M 2013 Dielectric fluid in inhomogeneous pulsed electric field *Phys. Rev.* E **87** 043004

[9] Korobeinikov S M, Melekhov A V and Besov A S 2002 Breakdown initiation in water with the aid of bubbles *High Temp.* **40** 652–9

[10] Mededovic S and Locke B R 2007 Primary chemical reactions in pulsed electrical discharge channels in water *J. Phys. D: Appl. Phys.* **40** 7734

[11] Locke B R and Thagard S M 2012 Analysis and review of chemical reactions and transport processes in pulsed electrical discharge plasma formed directly in liquid water *Plasma Chem. Plasma Process.* **32** 875–917

[12] Sponsel N 2024 Characteristics of pulsed electric breakdown in water-submerged gas bubbles *PhD Thesis* North Carolina State University

[13] Sponsel N L, Gershman S and Stapelmann K 2023 Electrical breakdown dynamics in an argon bubble submerged in conductive liquid for nanosecond pulsed discharges *J. Phys. D: Appl. Phys.* **56** 505202

[14] Stancampiano A, Gallingani T, Gherardi M, Machala Z, Maguire P, Colombo V, Pouvesle J -M and Robert E 2019 Plasma and aerosols: Challenges, opportunities and perspectives *Appl. Sci.* **9** 3861

[15] Nayak G, Meyer M, Oinuma G, Kushner M J and Bruggeman P J 2023 The transport dynamics of tens of micrometer-sized water droplets in rf atmospheric pressure glow discharges *Plasma Sources Sci. Technol.* **32** 045005

[16] Nayak G, Wang J, Li R, Aranzales D, Thagard S M and Bruggeman P J 2023 Non-OH-driven liquid-phase chemistry in water microdroplets *Plasma Process. Polym.* **20** 2200222

[17] Kruszelnicki J, Lietz A M and Kushner M J 2019 Atmospheric pressure plasma activation of water droplets *J. Phys. D: Appl. Phys.* **52** 355207

[18] Janda M, Stancampiano A, Natale F D and Machala Z 2025 Short review on plasma-aerosol interactions *Plasma Process. Polym.* **22** 2400275

[19] Green F J and Hasan M I 2025 Kinetic processes of interfacial transport of reactive species across plasma-water interfaces: the effect of temperature *Phys. Chem. Chem. Phys.* **27** 2218–31
[20] Heirman P and Bogaerts A 2024 Critical comparison of interfacial boundary conditions in modelling plasma-liquid interaction *J. Phys. D: Appl. Phys.* **58** 085206
[21] Sander R 2023 Compilation of Henry's law constants (version 5.0. 0) for water as solvent *Atmospheric Chem. Phys.* **23** 10901–2440
[22] Myers B, Dogariu A, Beeler B, Byrom L, Yatom S and Stapelmann K 2025 Imaging solvated oxygen atoms with a femtosecond laser *Nat. Commun.* (accepted)
[23] Sgonina K, Bruno G, Wyprich S, Wende K and Benedikt J 2021 Reactions of plasma-generated atomic oxygen at the surface of aqueous phenol solution: Experimental and modeling study *J. Appl. Phys.* **130** 043303
[24] Hassan M E, Janda M and Machala Z 2021 Transport of gaseous hydrogen peroxide and ozone into bulk water vs. electrosprayed aerosol *Water* **13** 182
[25] Verlackt C C W, Boxem W V and Bogaerts A 2018 Transport and accumulation of plasma generated species in aqueous solution *Phys. Chem. Chem. Phys.* **20** 6845–59
[26] Semenov I L, Weltmann K D and Loffhagen D 2019 Modelling of the transport phenomena for an atmospheric-pressure plasma jet in contact with liquid *J. Phys. D: Appl. Phys.* **52** 315203
[27] Silsby J A, Simon S, Walsh J L and Hasan M I 2021 The influence of gas-liquid interfacial transport theory on numerical modelling of plasma activation of water *Plasma Chem. Plasma Process.* **41** 1363–80
[28] Poggemann H F, Schüttler S, Schöne A L, Jeß E, Schücke L, Jacob T, Gibson A R, Golda J and Jung C 2025 Transportation behaviour of oh and h2o2 in plasma-treated water *J. Phys. D: Appl. Phys.* **58** 135208
[29] Yusupov M, Neyts E C, Simon P, Berdiyorov G, Snoeckx R, Van Duin A C T and Bogaerts A 2013 Reactive molecular dynamics simulations of oxygen species in a liquid water layer of interest for plasma medicine *J. Phys. D: Appl. Phys.* **47** 025205
[30] Rumbach P, Bartels D M and Go D B 2018 The penetration and concentration of solvated electrons and hydroxyl radicals at a plasma-liquid interface *Plasma Sources Sci. Technol.* **27** 115013
[31] Campo M G and Grigera J R 2005 Classical molecular-dynamics simulation of the hydroxyl radical in water *J. Chem. Phys.* **12** 0584507
[32] Denicola A, Souza J M, Radi R and Lissi E 1996 Nitric oxide diffusion in membranes determined by fluorescence quenching *Arch. Biochem. Biophys.* **328** 208–12
[33] Wise D L and Houghton G J C E S 1968 Diffusion coefficients of neon, krypton, xenon, carbon monoxide and nitric oxide in water at 10-60 °C *Chem. Eng. Sci.* **23** 231211–6
[34] Zhou Z, Todd B D, Travis K P and Sadus R J 2005 A molecular dynamics study of nitric oxide in water: Diffusion and structure *J. Chem. Phys.* **123** 054505
[35] Krynicki K, Green C D and Sawyer D W 1978 Pressure and temperature dependence of self-diffusion in water *Faraday Discuss. Chem. Soc.* **66** 199–208
[36] Debye P J F 1929 *Polar Molecules* (Dover)
[37] Alvarez F, Arbe A and Colmenero J 2023 The Debye's model for the dielectric relaxation of liquid water and the role of cross-dipolar correlations. a MD-simulations study *J. Chem. Phys.* **159** 134505

[38] Elg D T, Delgado H E, Martin D C, Sankaran R M, Rumbach P, Bartels D M and Go D B 2021 Recent advances in understanding the role of solvated electrons at the plasma-liquid interface of solution-based gas discharges *Spectrochim Acta* B **186** 106307

[39] Rumbach P, Bartels D M, Sankaran R M and Go D B 2015 The solvation of electrons by an atmospheric-pressure plasma *Nat. Commun.* **6** 7248

[40] Martin D C, Bartels D M, Rumbach P and Go D B 2021 Experimental confirmation of solvated electron concentration and penetration scaling at a plasma-liquid interface *Plasma Sources Sci. Technol.* **30** 03LT01

[41] Sutmann G 1998 Structure formation and dynamics of water in strong external electric fields *J. Electroanal. Chem.* **450** 289–302

[42] Okuno Y, Minagawa M, Matsumoto H and Tanioka A 2009 Simulation study on the influence of an electric field on water evaporation *J. Mol. Struct.* **904** 83–90

[43] Minagawa Y, Shirai N, Uchida S and Tochikubo F 2013 Analysis of effect of ion irradiation to liquid surface on water molecule kinetics by classical molecular dynamics simulation *Jpn. J. Appl. Phys.* **53** 010210

[44] Vanraes P and Bogaerts A 2021 The essential role of the plasma sheath in plasma-liquid interaction and its applications–a perspective *J. Appl. Phys.* **129** 220901

[45] Rumbach P, Clarke J P and Go D B 2017 Electrostatic Debye layer formed at a plasma-liquid interface *Phys. Rev.* E **95** 053203

[46] Tokman M, Lee J H, Levine Z A, Ho M C, Colvin M E and Vernier P T 2013 Electric field-driven water dipoles: nanoscale architecture of electroporation *PLoS One* **8** e61111

[47] Buraschi M, Horsfield A P and Cucinotta C S 2024 Revealing interface polarization effects on the electrical double layer with efficient open boundary simulations under potential control *J. Phys. Chem. Lett.* **15** 4872–9

[48] Delgado H E, Elg D T, Bartels D M, Rumbach P and Go D B 2020 Chemical analysis of secondary electron emission from a water cathode at the interface with a nonthermal plasma *Langmuir* **36** 1156–64

[49] Ma J, Wang F and Mostafavi M 2018 Ultrafast chemistry of water radical cation, $H_2O^{\bullet+}$, in aqueous solutions *Molecules* **23** 244

[50] Bruggeman P J, Frontiera R R, Kortshagen U, Kushner M J, Linic S, Schatz G C, Andaraarachchi H, Chaudhuri S, Chen H -T and Clay C D 2025 Advances in plasma-driven solution electrochemistry *J. Chem. Phys.* **162** 071001

[51] George N K, Wols B A and Gernjak W 2024 Development and validation of a simplified kinetic model for predicting organic micropollutant degradation in vacuum ultraviolet water treatment *Chem. Eng. J.* **499** 155998

[52] Stapelmann K, Gershman S and Miller V 2024 Plasma-liquid interactions in the presence of organic matter–a perspective *J. Appl. Phys.* **135** 160901

[53] Sander M U, Luther K and Troe J 1993 On the photoionization mechanism of liquid water *Ber. Bunsenges. Phys. Chem.* **97** 953–60

[54] Elles C G, Shkrob I A, Crowell R A and Bradforth S E 2007 Excited state dynamics of liquid water: Insight from the dissociation reaction following two-photon excitation *J. Chem. Phys.* **126** 164503

[55] Attri P, Kim Y H, Park D H, Park J H, Hong Y J, Uhm H S, Kim K N, Fridman A and Choi E H 2015 Generation mechanism of hydroxyl radical species and its lifetime prediction during the plasma-initiated ultraviolet (uv) photolysis *Sci. Rep.* **5** 9332

[56] IUPAC 2025 'Geminate recombination' *IUPAC Compendium of Chemical Terminology* 5th edn (International Union of Pure and Applied Chemistry) 2025
[57] Gonzalez M G, Oliveros E, Wörner M and Braun A M 2004 Vacuum-ultraviolet photolysis of aqueous reaction systems *J. Photochem. Photobiol.* **5** 225–46
[58] Yang Y and Pignatello J J 2017 Participation of the halogens in photochemical reactions in natural and treated waters *Molecules* **22** 1684
[59] Kalmár J, Dóka E, Lente G and Fábián I 2014 Aqueous photochemical reactions of chloride, bromide, and iodide ions in a diode-array spectrophotometer. autoinhibition in the photolysis of iodide ions *Dalton Trans.* **43** 4862–70
[60] Harada Y, Masuda S and Ozaki H 1997 Electron spectroscopy using metastable atoms as probes for solid surfaces *Chem. Rev.* **97** 1897–952
[61] Berendsen C W J, van Veldhuizen E M, Kroesen G M W and Darhuber A A 2014 Marangoni flows induced by atmospheric-pressure plasma jets *J. Phys. D: Appl. Phys.* **48** 025203
[62] Kawasaki T, Kamasaki M, Takeuchi N and Mitsugi F 2021 Effects of initial surfactant concentration on plasma-induced liquid flows *J. Appl. Phys.* **130** 243303
[63] Thagard S M, Stratton G R, Vasilev M, Conlon P and Bohl D 2018 An experimental investigation of the liquid flow induced by a pulsed electrical discharge plasma *Plasma Chem. Plasma Process.* **38** 719–41
[64] Tochikubo F, Aoki T, Shirai N and Uchida S 2017 Characteristics of liquid flow induced by atmospheric-pressure dc glow discharge in contact with liquid *Jpn. J. Appl. Phys.* **56** 046201
[65] Holgate J T, Coppins M and Allen J E 2019 Simulated dynamics of a plasma-sheath-liquid interface *New J. Phys.* **21** 063002
[66] Lai J, Petrov V and Foster J E 2018 Understanding plasma-liquid interface instabilities using particle image velocimetry and shadowgraphy imaging methods *IEEE Trans. Plasma Sci.* **46** 875–81
[67] Kogelheide F 2021 Characterisation of a dielectric barrier discharge and its impact on liquids and biological systems *PhD Thesis* Ruhr-Universität Bochum, Bochum
[68] Yang Z, Kovach Y and Foster J 2021 Visualization and analysis of coupling between plasmas self-organization and plasma-induced fluid circulation in 1 atm dc glows with liquid anode *J. Appl. Phys.* **129** 163303
[69] Kawamoto H and Umezu S 2005 Electrohydrodynamic deformation of water surface in a metal pin to water plate corona discharge system *J. Phys. D: Appl. Phys.* **38** 887
[70] Bruggeman P, Graham L, Degroote J, Vierendeels J and Leys C 2007 Water surface deformation in strong electrical fields and its influence on electrical breakdown in a metal pin-water electrode system *J. Phys. D: Appl. Phys.* **40** 4779
[71] Dickenson A, Walsh J L and Hasan M I 2021 Electromechanical coupling mechanisms at a plasma-liquid interface *J. Appl. Phys.* **129** 213301
[72] Park S, Choe W, Lee H, Park J Y, Kim J, Moon S Y and Cvelbar U 2021 Stabilization of liquid instabilities with ionized gas jets *Nature* **592** 49–53
[73] Bordbar M, Hajisharifi K, Mehdian H, Hasanbeigi A and Robert E 2024 Dependence of ehd instability of plasma/liquid interface on liquid conductivity *Appl. Phys. Lett.* **124**
[74] Dubrovski O, Yang J, Veloso F, Go D B, Chang H C and Rumbach P 2024 Noncapillary wave dynamics due to interfacial coupling with plasma patterns at a liquid surface *Phys. Rev. Lett.* **133** 105301

[75] Doyle S J, Dias T C, Meyer M and Kushner M J 2025 Controlling plasma produced fluxes to liquid surfaces by acoustic structuring: applications to plasma driven solution electrochemistry *Plasma Sources Sci. Technol.* **34** 035004

[76] Albertson T G and Troian S M 2019 Electrified cone formation in perfectly conducting viscous liquids: Self-similar growth irrespective of reynolds number *Phys. Fluids* **31** 102103

[77] Gilbert W 1958 *De magnete* (Dover)

[78] Gray S 1732 II. A letter concerning the electricity of water, from Mr. Stephen Gray to Cromwell Mortimer, MD Secr. R. S *Phil. Trans. R. Soc.* **37** 227–60

[79] Taylor G I 1964 Disintegration of water drops in an electric field *Phil. Trans. R. Soc.* A **280** 383–97

[80] Melcher J R and Smith C V 1969 Electrohydrodynamic charge relaxation and interfacial perpendicular field instability *Phys. Fluids* **12** 778–90

[81] Jirsák J, Moučka F, Škvor J and Nezbeda I 2015 Aqueous electrolyte surfaces in strong electric fields: molecular insight into nanoscale jets and bridges *Mol. Phys.* **113** 848–53

[82] Luedtke W D, Landman U, Chiu Y H, Levandier D J, Dressler R A, Sok S and Gordon M S 2008 Nanojets, electrospray, and ion field evaporation: Molecular dynamics simulations and laboratory experiments *J. Phys. Chem.* A **112** 9628–49

[83] Masson A and Nykyri K 2018 Kelvin-Helmholtz instability: Lessons learned and ways forward *Space Sci. Rev.* **214** 71

[84] Lietz A M, Johnsen E and Kushner M J 2017 Plasma-induced flow instabilities in atmospheric pressure plasma jets *Appl. Phys. Lett.* **111** 114101

[85] Kondeti V S S K and Bruggeman P J 2020 The interaction of an atmospheric pressure plasma jet with liquid water: Dimple dynamics and its impact on crystal violet decomposition *J. Phys. D: Appl. Phys.* **54** 045204

[86] Barron M A, Medina D Y and Reyes J 2021 Interaction between a liquid surface and an impinging gas jet *World J. Eng. Technol.* **9** 793–803

[87] Mordasov M M, Savenkov A P and Chechetov K E 2016 Method for analyzing the gas jet impinging on a liquid surface *Tech. Phys.* **61** 659–68

[88] Meyer M, Nayak G, Bruggeman P J and Kushner M J 2022 Sheath formation around a dielectric droplet in a he atmospheric pressure plasma *J. Appl. Phys.* **132** 083303

IOP Publishing

Katharina Stapelmann

Chapter 6

Application examples

Atmospheric pressure plasmas at interfaces are powerful precisely because they couple fields, charges, photons, and chemically active species to surfaces and liquids whose properties evolve in time. The earlier chapters laid out the ingredients: interfacial adsorption and solvation, transport and renewal, electrochemical pathways including solvated electrons and Faradaic steps, photon/metastable-driven reactions, and plasma-induced fluid motion. This chapter brings those threads together in two application examples that stress-test the full framework: plasma–bio interactions and plasma-enabled water treatment. In both, outcomes are ultimately decided at or within a few hundred nanometers of the interface, yet they depend on device waveforms and geometry, liquid composition and conductivity, mixing and renewal, and the characteristic timescales that set whether a system behaves as dielectric or conductive (cf Maxwell relaxation), diffusion- or reaction-limited.

In plasma–bio systems, the 'sample' is alive and actively reshapes its chemical and electrical environment: media composition sets boundary conditions, cells consume and generate species, and viability changes dielectric properties and thus the plasma circuit itself. Dose cannot be a single scalar; it spans input, delivered, effective, and biological response layers that depend on both source and biology. In pollutant removal, the target chemistry is often simpler but the transport challenge is starker: performance hinges on placing contaminants at the reaction zone (the interface) long enough to meet the right agents: oxidants for many organics, reductants and VUV/e_{aq} for perfluoroalkyl and polyfluoroalkyl substances (PFAS), without over-treating or quenching chemistry in the bulk. Across both applications, the same levers recur: maximize the useful interfacial area, control renewal and residence time, tune gas and waveform to sculpt the species mix, and let material properties (permittivity, conductivity, surface activity) guide whether reactions are surface-localized or bulk-relevant. The case studies that follow translate these principles into practice and, just as importantly, show how to diagnose when transport, not production, sets your limits.

doi:10.1088/978-0-7503-6254-2ch6

6.1 Application I—Plasma–bio interactions

At-a-glance
Liquids and living targets co-define both chemistry and circuit: media composition, cell density/viability, and timescales (e.g., τ_M) shape what is delivered and how it is interpreted biologically. 'Dose' spans input → delivered → effective → response layers, with feedbacks that make cell state a control parameter.

Plasma–bio interactions are inherently multi-faceted, as they combine plasma physics, liquid chemistry, and biological complexity into one experimental system. Unlike studies of plasma–solid or plasma–liquid interfaces in isolation, biological applications combine both interfaces into a highly complex, coupled system. Here, plasmas encounter not only liquids of varying composition but also living or non-living biological material. This section introduces experimental setups, liquid environments, and biological systems commonly used in plasma–bio research; it is by no means exhaustive, as it could probably fill a whole textbook by itself. Parts of this section are based on a perspective paper by the author [1].

A variety of plasma sources have been employed to study plasma interactions with biological targets, with DBDs and plasma jets being the most common. Direct treatment systems, where the active plasma region is in contact with cells or tissues, allow exposure to electric fields, charged particles, metastables, and short-lived species, whereas indirect systems, such as surface DBDs and remote jets allow exposure to longer-lived species (cf sections 3.1.3, 4.4, and 5.6).

The liquids used in these studies are equally diverse. Simple aqueous buffers such as phosphate buffered saline (PBS) or cell culture medium (e.g., RPMI, DMEM, with and without serum) introduce very different chemistries. Their ionic strength, pH buffering capacity, and presence of biomolecules not only shape the plasma-induced reactions but also influence transport, conductivity, and interfacial properties. The liquid medium thus becomes more than a passive layer: it participates in the chemistry and serves as the boundary condition for the plasma–liquid interface, with the potential to alter the plasma properties.

Experimental plasma–bio systems have progressively increased in complexity: from studies on biomolecules [2, 3] and proteins [4, 5], to bacteria [6, 7] and cell cultures [8, 9], tissue slices [10, 11], and eventually whole organisms [12, 13]. For *in vitro* studies, setups often involve multi-well plates with small liquid volumes. Even with a focus on *in vitro* investigations, there is a large variety of different cell lines used, ranging from keratinocytes for wound healing studies to a vast variety of cancer cell lines. As shown in Mohamed *et al* and discussed more in section 6.1.3, two different leukemic cancer cell lines exhibited very different responses to plasma treatment [8]. Given the complexity of the biological targets, different outcomes and results can be expected depending on the plasma type and the biological targets used. However, there are some general observations and specific examples that can help understanding the complex interactions plasma–bio systems experience, discussed in the following sections.

6.1.1 The role of the liquid medium

The liquid medium is both the transmitter and the modulator of plasma activity. Plasma-generated species must cross the plasma–liquid interface, where transport is influenced by interfacial gradients, transport processes, and possibly electric fields (cf chapter 5). Once inside the liquid, species lifetimes and diffusion depths define how far reactivity extends. Importantly, liquids with high organic content (e.g., cell culture medium) undergo complex secondary chemistry that diverges strongly from neat water models [1]. Cell culture medium consists of multiple components, including glucose, amino acids, vitamins, inorganic salts, and a buffering system. Some media contain phenol red as a pH indicator; phenol red, however, acts as a radical scavenger of plasma-generated reactive species. Most media are furthermore supplemented with fetal bovine serum and antibiotics. The medium and supplements chosen for the experiments can strongly influence results, as discussed in [9]. Different formulations (e.g., DMEM, RPMI, MEM) vary in glucose, amino acids, and buffers, which partly explains the variability between studies in plasma–cell culture experiments [9]. The buffering capacity of the medium can affect the pH shifts during plasma treatment. The antioxidant capacity of the medium, arising from radical scavengers such as phenol red, amino acids (e.g., cysteine or methionine), and serum proteins, can drastically alter the availability of plasma-generated species. In fact, many amino acids, vitamins, and proteins act as intrinsic radical scavengers. In this sense, the cell culture medium actively participates in the chemistry rather than serving as a passive solvent. Furthermore, the ionic strength of the medium influences the conductivity, impacting the electrical boundary conditions of the plasma liquid interface (cf section 5.3). Liquids in direct contact with the plasma can thus be regarded as electrical components. Their conductivity defines the current closure while their permittivity defines field distribution [1]. Thus, the liquid medium cannot be viewed as a neutral environment but as an active electrical component.

6.1.2 Complex chemistry in the presence of biological material

Introducing biological material into the plasma–liquid system dramatically alters both plasma and liquid chemistry. Proteins, lipids, and nucleic acids present in media or released from cells provide abundant reaction partners for plasma-generated species. For example, cysteine consumes H_2O_2 after an initial chemical reaction with a more reactive species such as O or OH [14]. In contrast, aromatic ring structures such as phenol [15], HTA, and DMPO [16] can *increase* H_2O_2 concentrations. The likely mechanism involves atomic oxygen abstracting a hydrogen atom from the ring, which then reacts with dissolved O_2 to form HO_2 or OH radicals. These precursors recombine, yielding additional H_2O_2 [16].

Chemical modifications of biomolecules therefore reshape the reactive landscape of the liquid by shifting species concentrations and equilibrium pathways. Such interactions should not be reduced to simple 'scavenging' but understood as dynamic processes in which organic matter drives secondary reactions, alters equilibrium species densities, and even feeds back into plasma transport across the interface [1]. Computational work by Lietz and Kushner illustrates this feedback clearly: in the presence of peptidoglycan, liquid-phase ozone is consumed, sustaining uptake from the gas phase. This decreases gas-phase ozone concentrations, reduces electron attachment, and increases electron density by 4.7% [17]. The higher electron density in turn enhances the production of N, O, OH, and $O_2(^1\Delta_g)$. Thus, the presence of biomaterial can directly reshape gas-phase chemistry and plasma behavior.

Beyond altering equilibria, plasma-generated reactivity can be *transferred* to the biomolecules themselves. In one study, albumin, glucose, cholesterol, and arachidonic acid were plasma-modified and then tested *in vitro* and *in vivo* [18]. *In vitro*, mutagenesis increased when cells were cultured with these modified biomolecules, whereas *in vivo* experiments in insect larvae showed no toxicity, suggesting that higher organisms may detoxify reactive components. In another study, plasma-modified cysteine was introduced to human keratinocyte cultures [19]. At physiological concentrations, no effects were observed; however, at artificially high concentrations, decreased proliferation, migration, and lactate secretion were detected, along with enhanced cytostatic and cytotoxic effects. The authors proposed that modified cysteine may act as a 'preserving agent of the chemical energy of plasma-derived species.'

Together, these examples demonstrate that biological material in plasma–liquid systems does not merely absorb plasma reactivity but actively participates in, sustains, and redistributes it, with consequences ranging from altered plasma–liquid density equilibria to biological effects.

6.1.3 Interaction of plasma with living cells

Plasma–cell interactions are inherently bidirectional. The discharge delivers short-lived and long-lived reactive oxygen and nitrogen species (RONS), electric fields, photons, metastables, and charged species into a cell culture, while cells dynamically reshape this milieu through enzymatic detoxification, metabolite release, membrane remodeling, and changes to bulk dielectric properties. As a result, 'dose' is co-determined by source settings *and* cell state: antioxidant capacity, metabolic activity, and even viability feed back into both the chemistry at the interface and the electrical behavior of the source.

Mohamed *et al* demonstrated that different leukemic cell lines (Jurkat T lymphocytes versus THP-1 monocytes) modulate plasma-induced chemistry in distinct ways. While both cell types depleted hydrogen peroxide from the medium, only Jurkat cells showed strong mitochondrial superoxide responses and immunogenic marker expression. In contrast, THP-1 cells neutralized hydrogen peroxide rapidly and exhibited higher resistance to plasma cytotoxicity [8]. These results highlight a key point: biological systems are not simply exposed to plasma chemistry but actively reshape it. Depending on their enzymatic machinery and antioxidant capacity, cells alter the balance of RONS in the surrounding liquid, feeding back into both plasma chemistry and biological outcomes. Cells may act passively, as a target for reactive species, succumbing to oxidative stress, DNA damage, and membrane disruption. However, they can also actively respond, adapting their redox metabolism, releasing metabolites, and modifying their immediate environment. Mohamed *et al* showed that leukemic cells not only modulate RONS levels but also display immunogenic markers (e.g., calreticulin, HSP70/90), stimulating phagocytosis by antigen-presenting cells even when direct cytotoxicity was minimal [8].

In Sutter *et al*, a floating electrode dielectric barrier discharge (DBD) was specifically analyzed for its electrical properties when biological targets were included, revealing that cell suspensions themselves affect discharge behavior [20]. Even viable versus dead or lysed cells change the dielectric properties of the suspension differently, altering the capacitance of the plasma circuit. These observations emphasize that biological targets are not passive recipients of plasma but integral components of the plasma circuit. Moreover, liquids act as conducting media. Sutter *et al* demonstrated that the dielectric properties of liquids with or without cells directly influence circuit capacitance and energy deposition in DBDs, creating feedback loops between the plasma source and the biological target [20]. Plasma modifies the cells, and the cells in turn modify plasma discharge properties.

From the plasma's perspective, a cell suspension behaves as a slab with a complex admittance Y_A per area:

$$Y_A(\omega) = \sigma_{eff} + j\ \omega\varepsilon_0\varepsilon_{eff} \tag{6.1}$$

so its impedance is $Z_A(\omega) = 1/Y_A$. Thus, higher conductivity σ_{eff} (e.g., serum proteins, ion leakage from lysed cells) reduces $|Z_A|$ and draws more current, shifting voltage partitioning away from the suspension and altering energy deposition in DBD circuits. Higher permittivity ε_{eff} increases C_{susp}:

$$C_{susp} = \frac{\varepsilon_0\varepsilon_{eff}A}{d} \tag{6.2}$$

strengthening capacitive coupling and modifying sheath dynamics at the interface. The relevant timescale is the Maxwell relaxation time $\tau_M = \varepsilon_0\varepsilon_{eff}/\sigma_{eff}$: for excitations with rise time or period $\ll\tau_M$ the suspension responds capacitively (dielectric), whereas for $\gg\tau_M$ it responds conductively (ohmic). Because ε_{eff} and σ_{eff} depend on cell viability and packing (membrane-encapsulated cells versus lysed cytoplasm), living samples can evolve during treatment, feeding back into discharge current, sheath voltage, and ultimately the delivered plasma 'dose.' Consistent with this view, Sutter *et al* observed that viable versus dead/lysed cells shift σ_{eff} and ε_{eff}, thereby changing $Z_A(\omega)$ and the effective circuit capacitance seen by the floating electrode DBD source.

In practice, plasma 'dose' emerges from a coupled source–liquid–cell system; cellular biochemistry and suspension electrodynamics co-determine both what is delivered and how it is interpreted biologically.

6.1.3.1 Why 'dose' is difficult to define in plasma–cell systems

Unlike ionizing radiation or small-molecule drugs, atmospheric pressure plasmas deliver a cocktail of agents (short-/long-lived RONS, photons, fields, charge) into a *coupled* source–liquid–cell system whose properties evolve during exposure. Timescales further complicate interpretation. The exposure happens on the order of seconds, chemistry can continue over seconds to minutes and first cellular responses come into play (e.g., calcium release). The biological outcomes, however, can only be observed hours (for cell culture) to days or weeks (whole organisms) later. As a result, a single scalar 'dose' cannot capture (i) multi-component inputs, (ii) transport and consumption in the liquid, (iii) cell-dependent feedbacks, and (iv) time- and space-dependent delivery. It is therefore useful to distinguish four layers:

1. **Source (input) dose**—What the device outputs by its settings (waveform, voltage/current, pulse width/frequency, gas/flow, gap, exposure time). Useful electrical surrogates are power (W), energy per pulse (J), energy density ($J\,m^{-2}$) for surfaces, or energy per liquid volume ($J\,L^{-1}$).
2. **Delivered dose at the interface/liquid**—What reaches and accumulates at the sample: charge transferred (C), energy deposited (J), photon fluence (photons cm^{-2}), and time-integrated chemical exposures (e.g., $\int c_{H_2O_2}\,dt$; interfacial fluxes $\int \Gamma_i\,dt$ in $mol\,m^{-2}$). This already depends on medium composition and geometry.
3. **Effective dose at the cell**—What arrives at the cell surface or intracellularly. It depends on diffusion, convection, enzymatic consumption (catalase/GPx/SOD), cell density and viability, and microenvironment (2D monolayer versus 3D spheroid/pellet). In practice this is rarely measured directly and is highly context-dependent.
4. **Biological response (effect)**—Readouts such as viability, apoptosis/necrosis, mito-ROS, and ICD/DAMPs (CRT, HSP70/90). These are *consequences*, not dose itself, and are nonlinearly related to the upstream layers.

Why a single number fails: (a) *multidimensionality* (different agents act synergistically/antagonistically), (b) *path dependence* (pulse structure/duty cycle and sequence matter), (c) *feedback* (cells and liquids change dielectric properties and RONS sinks/sources during exposure), (d) *spatial heterogeneity* (nonuniform fields/flows, well-edge effects, 3D gradients), and (e) *measurement artifacts* (aromatic probes or scavengers can alter the chemistry).

6.2 Application II—Removal of pollutants from water

At-a-glance

Most removal rates are transport-governed: maximize interfacial area and convective renewal to present contaminants at the reaction zone. Oxidants clear many organics, while PFAS demand interfacial reduction pathways (e^-_{aq}, VUV) and devices that exploit their surface activity.

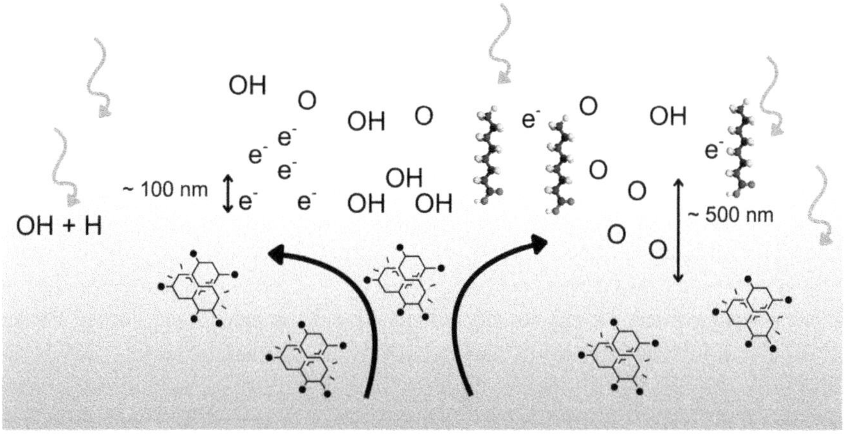

Figure 6.1. Conceptual overview of oxidizing and non-oxidizing species at the plasma–liquid interface. A representative surfactant is located at the interface and a representative organic pollutant is in the bulk liquid. The black arrows indicate a flow pattern that may have been introduced by the discharge. Typical penetration depths of OH and e^- are given as ~100 nm [21] and of O as ~500 nm [22] as upper boundaries. The orange arrows indicate UV/VUV photons.

Plasmas in or in contact with water have been investigated for pollutant removal from water, originally assuming that it is a novel advanced oxidation process (AOP). Recently, the non-oxidizing species have attracted more attention, e.g., electrons, metastables, and (V)UV photons. Many organic pollutants are readily oxidized by plasma-generated OH, O_3, O, etc. However, the forever chemicals PFAS are non-oxidizable due to their strong C–F bond. While many organic pollutants would be distributed in the bulk liquid, PFAS are surfactants and thus localized at the plasma–liquid interface. For the removal of persistent organic compounds, different strategies might need to be used than for PFAS removal. It can roughly be separated into bulk liquid versus interface chemistry and oxidation versus reduction of the pollutants. Figure 6.1 shows a conceptual overview of oxidizing and non-oxidizing species at the plasma–liquid interface. A representative surfactant is located at the interface (of course, surfactants are not exclusively located at the interface) and exposed to all components of the discharge (assuming a direct discharge), including electrons. A representative organic pollutant is in the bulk liquid. Plasma-induced flow or active mixing of the liquid (indicated with the black arrows) can help to bring the pollutant closer to the surface to provide exposure to more components of the plasma. The upper limit penetration depths of OH and electrons have been shown to be around ~100 nm [21] and for atomic oxygen ~500 nm [22]. Thus, it is important to consider mixing, described in more detail below (6.2.2).

6.2.1 The role of the pollutant

As mentioned above, different pollutants require different strategies for removal. Some organic compounds can easily be oxidized, while PFAS or oxyanions, for example, require reducing reactions. PFAS are challenging to destroy due to their strong C–F bonds and low reactivity toward common oxidants as they are lacking C–H bonds or C–C π systems that are usually attached by OH or ozone [23]. While the exact pathways are still under investigation, a mechanistic picture starts to emerge: as surfactants, long-chain PFAS enrich at gas–liquid interfaces [24], positioning them where the discharge delivers electrons and VUV photons. Reductive steps initiated by hydrated electrons (e_{aq}^{-}) open the first C–F bonds [25]. VUV can contribute both directly and indirectly. Direct photolysis of certain PFAS occurs at 185 nm [26] (and, for some, at 222 nm [27]). Indirectly, VUV photolyzes water to generate e_{aq}^{-}. However, both, e_{aq}^{-} and VUV do not penetrate the liquid well, so these processes are likely confined to the interface. As many PFAS are surfactants, it has also been hypothesized that gas-phase electrons and VUV may interact with the PFAS molecule directly before entering the liquid phase [28]. Because the reactivity takes place primarily at the plasma–liquid interface, plasma-based PFAS removal appears to be less sensitive to the presence of organic and inorganic co-contaminants, unlike other treatment processes [28].

Vasilev *et al* [29] reported two kinetic regimes for plasma-induced degradation of organic contaminants: (i) production-limited, where the reactive species production limits the degradation for high solute concentrations, and (ii) mass-transport-limited, where enough reactive species are produced but do not reach the contaminant, e.g., in low solute concentrations. Thus, it is important for reactor design to consider initial contaminant concentrations to address these limitations. For the mass-transport-limited regime, gas bubbling, droplets/aerosols, flowing film or rotating disk setups can be used to improve the efficiency, described in the next section.

6.2.2 Current gas–liquid plasma reactor concepts for the removal of pollutants

The limited but growing literature indicates that, for most contaminants, removal rates are governed more by transport to the plasma–liquid interface than by the intrinsic production rate of reactive species [30–33]. High performance is typically realized by (i) maximizing plasma–liquid interfacial area through electrode geometry and power–coupling choices, and (ii) operating in a convection-limited regime with rapid renewal of liquid at the interface which can be achieved via bubbling, spraying, or engineered mixing patterns (e.g., baffles) [34]. In contrast, diffusion-limited operation keeps the liquid in contact with the discharge for too long, promoting over-treatment and mass-transfer bottlenecks that diminish overall effectiveness [29, 35].

Surfactant-like contaminants (e.g., dyes, long-chain PFAS) preferentially accumulate at the interface; even modest changes in mixing can raise their Gibbs surface excess and accelerate removal [36]. For these species, apparent rates often increase with bulk concentration (pseudo-first-order behavior). Destruction becomes limited by the generation of reactive species when a certain threshold concentration is reached. This transition can be shifted to higher concentrations by increasing deposited power and potentially by tuning other operating parameters [37].

By contrast, many pharmaceuticals and pesticides exhibit low interfacial enrichment and remain predominantly in bulk liquid, reducing their diffusive flux to the reaction zone; solvated reactive species may recombine before encountering them, yielding slower removal. For such non-surfactants, reactors must maintain high fluid renewal rates so that the maximum attainable concentration is persistently presented at the interface. Spinning-disk and bubbling configurations have been especially effective in achieving this condition [35, 37]. A recent opinion piece nicely describes how to design reactors in plasma–liquid systems for wastewater treatment and is recommended for further reading [38]. In [39], example applications of plasma for the removal of various emerging contaminants are reviewed and the integration of plasma systems as a pre- and post-treatment process with other treatment technologies are discussed. An extensive review article from 2021 describes the different device concepts and outcomes for the removal of organic micropollutants with a focus on organic dyes, pharmaceuticals, and pesticides [40]. A few concepts focused on PFAS degradation will be described in the following as examples.

Several discharge concepts have been studied for the degradation of PFAS. As the longer-chained PFAS molecules are surface-active, many of the devices are designed to optimize the surface-to-volume ratio albeit with different concepts.

A continuous flowing gas–liquid film reactor that utilizes nanosecond pulsed power has shown a frequency-dependent reduction of perfluorooctanoic acid (PFOA) [41]. The energy yield increased with higher initial PFOA concentrations following a power law. When the liquid effluent was recycled and exposed to the plasma multiple times, PFOA was undetectable after the second pass. An analysis of the byproducts showed that the concentration of shorter-chain PFAS increased after plasma exposure, supporting the idea of stepwise degradation of PFAS. The authors note that lower-chain PFAS has low to no surface activity and that it is important to study ways to enhance the degradation of lower-chain PFAS to achieve complete degradation and mineralization [41]. The same point is reinforced in [38]. Furthermore, a two-step process is suggested with the first intended to concentrate the contaminant with ion exchange or reverse osmosis to allow more efficient treatment of reasonable volumes of highly concentrated waste [38].

The first pilot-scale plasma-based water treatment technology to be deployed on site for PFAS removal on an Air Force Base was introduced in [42]. The process owes its efficiency to the generation of plasma in argon gas above the liquid surface (argon is being recycled) and to argon gas bubbling via diffusers to transport the PFAS to the plasma–liquid interface. The reactor is composed of two identical setups with six microbubble diffuser and six high voltage sawblade electrodes per reactor. Each setup is individually driven by a HV power supply with a charging capacitor and a rotating spark gap. Long-chain perfluoroalkyl acids (PFAAs, fluorocarbon chain $\geqslant 6$) and PFAS precursors were reduced by more than 90% in a single cycle. Short-chain PFAAs (flurocarbon chain $\leqslant 5$) were reduced to a lesser extent. This could be overcome by introducing a cationic surfactant (cetrimonium bromide) by enhancing the short-chain species' transport to the plasma–liquid interface. A detailed energy consumption and cost analysis revealed that the plasma-based water treatment appears to be one of the most effective treatment options for PFOA and PFOS [42].

Summary—chapter 6

- **Reaction dynamics:** Design for where reactions actually happen, typically within ~0.1–0.5 μm of the surface, and keep renewal high enough to avoid quenching or over-treating.
- **Match agent to target:** Use oxidants for many organics; for PFAS, prioritize interfacial reduction (e^-_{aq}, VUV) and architectures that exploit surface enrichment.
- **Let material properties steer coupling:** Liquid conductivity/permittivity (and thus τ_M) set whether the system behaves capacitively or ohmically and how voltage partitions.
- **Treat 'dose' as layered:** Report/monitor input, delivered, and effective doses separately; interpret biological outcomes as responses, not proxies for upstream layers.
- **Engineer transport on purpose:** Bubbling, sprays/films, rotating or spinning surfaces, and jet-induced flows are design levers that often outrun chemistry tweaks.
- **Diagnose your regime:** Look for production-limited versus transport-limited signatures (e.g., concentration scaling, residence-time effects) and tune power/area/renewal accordingly.
- **Beware hidden feedbacks:** Cells, solutes, and byproducts change dielectric and chemical sinks/sources during exposure—measure them and expect dynamics.

Reading recommendations

- There are several textbooks about plasma–bio interactions available for further reading. A more clinical perspective for plasma medicine is given in [43] and [44].
- A more redox biology point of view for plasma–bio interactions is provided in [45].
- Section 6.1 is partly based on a perspective paper that is recommended for further reading: [1].
- Reference [38] is recommended for further reading for reactor design for wastewater treatment.
- Several publications review the use of plasma for other pollutants than PFAS, the focus pollutant in this chapter: antibiotics [46], plasma as advanced oxidation process tool [47], describing the importance of mechanistic studies for organic pollutant removal [23], cylindrospermopsin as model pollutant in water [48], recalcitrant pharmaceuticals [49], persistent organic compounds [50], to name a few.

References

[1] Stapelmann K, Gershman S and Miller V 2024 Plasma-liquid interactions in the presence of organic matter–a perspective *J. Appl. Phys.* **135** 160901

[2] Takai E, Kitamura T, Kuwabara J, Ikawa S, Yoshizawa S, Shiraki K, Kawasaki H, Arakawa R and Kitano K 2014 Chemical modification of amino acids by atmospheric-pressure cold plasma in aqueous solution *J. Phys. D: Appl. Phys.* **47** 285403

[3] Kogelheide F, Kartaschew K, Strack M, Baldus S, Metzler-Nolte N, Havenith M, Awakowicz P, Stapelmann K and Lackmann J-W 2016 FTIR spectroscopy of cysteine as a ready-to-use method for the investigation of plasma-induced chemical modifications of macromolecules *J. Phys. D: Appl. Phys.* **49** 084004

[4] Klinkhammer C, Verlackt C, Śmiłowicz D, Kogelheide F, Bogaerts A, Metzler-Nolte N, Stapelmann K, Havenith M and Lackmann J-W 2017 Elucidation of plasma-induced chemical modifications on glutathione and glutathione disulphide *Sci. Rep.* **7** 13828

[5] Lackmann J W, Baldus S, Steinborn E, Edengeiser E, Kogelheide F, Langklotz S, Schneider S, Leichert L I O, Benedikt J and Awakowicz P 2015 A dielectric barrier discharge terminally inactivates rnase a by oxidizing sulfur-containing amino acids and breaking structural disulfide bonds *J. Phys. D: Appl. Phys.* **48** 494003

[6] Krewing M, Jarzina F, Dirks T, Schubert B, Benedikt J, Lackmann J-W and Bandow J-E 2019 Plasma-sensitive escherichia coli mutants reveal plasma resistance mechanisms *J. R. Soc. Interface* **16** 20180846

[7] Jablonowski H, A Ch Hänsch M, Dünnbier M, Wende K, Hammer M U, Weltmann K D, Reuter S and von Woedtke T 2015 Plasma jetas shielding gas impact on bacterial inactivation *Biointerphases* **10** 029506

[8] Mohamed H, Gebski E, Reyes R, Beane S, Wigdahl B, Krebs F C, Stapelmann K and Miller V 2021 Differential effect of non-thermal plasma rons on two human leukemic cell populations *Cancers* **13** 2437

[9] Biscop E, Lin A, Van Boxem W, Van Loenhout J, De Backer J, Deben C, Dewilde S, Smits E and Bogaerts A 2019 The influence of cell type and culture medium on determining cancer selectivity of cold atmospheric plasma treatment *Cancers* **11** 1287

[10] Thompson T P, Connelly A, Kelly S, Duncan R M, Maybin J A, McDonnell C, Melvage A, McClenaghan L A, Dedeloudi A and Lamprou D A 2025 Biomedical application of cold plasma: Navigating through plasma types and devices by antimicrobial effectiveness and tissue tolerance *Adv. Ther.* **8** 2400339

[11] Maisch T, Shimizu T, Li Y-F, Heinlin J, Karrer S, Morfill G and Zimmermann J-L 2012 Decolonisation of MRSA, *s. aureus* and *e.coli* by cold-atmospheric plasma using a porcine skin model *in vitro PLoS One* **7** e34610

[12] Stratmann B, Costea T C, Nolte C, Hiller J, Schmidt J, Reindel J, Masur K, Motz W, Timm J and Kerner W 2020 Effect of cold atmospheric plasma therapy vs standard therapy placebo on wound healing in patients with diabetic foot ulcers: a randomized clinical trial *JAMA Netw. Open* **3** e2010411

[13] Schmidt A, Singer D, Aden H, von Woedtke T and Bekeschus S 2024 Gas plasma exposure alters microcirculation and inflammation during wound healing in a diabetic mouse model *Antioxidants* **13** 68

[14] Stapelmann K, Myers B, Quesada M H, Lenker E and Ranieri P J 2021 Following O and OH in He/O_2 and He/H_2O gas mixtures–from the gas phase through the liquid phase to modifications on a biological sample *J. Phys. D: Appl. Phys.* **54** 434003

[15] Hefny M M, Pattyn C, Lukes P and Benedikt J 2016 Atmospheric plasma generates oxygen atoms as oxidizing species in aqueous solutions *J. Phys. D: Appl. Phys.* **49** 404002

[16] Myers B, Ranieri P, Smirnova T, Hewitt P, Peterson D, Quesada M H, Lenker E and Stapelmann K 2021 Measuring plasma-generated· OH and O atoms in liquid using epr spectroscopy and the non-selectivity of the HTA assay *J. Phys. D: Appl. Phys.* **54** 145202

[17] Lietz A M and Kushner M J 2016 Air plasma treatment of liquid covered tissue: long timescale chemistry *J. Phys. D: Appl. Phys.* **49** 425204

[18] Heslin C, Boehm D, Gilmore B F, Megaw J, Freeman T A, Hickok N J, Cullen P J and Bourke P 2021 Biomolecules as model indicators of *in vitro* and *in vivo* cold plasma safety *Front. Phys.* **8** 613046

[19] Heusler T, Bruno G, Bekeschus S, Lackmann J W, von Woedtke T and Wende K 2019 Can the effect of cold physical plasma-derived oxidants be transported via thiol group oxidation? *Clin. Plasma Med.* **14** 100086

[20] Sutter J, Brettschneider J, Mamchur S, Krebs F, Gershman S and Miller V 2023 Inclusion of biological targets in the analysis of electrical characteristics of non-thermal plasma discharge *Plasma* **6** 577–91

[21] Rumbach P, Bartels D M and Go D B 2018 The penetration and concentration of solvated electrons and hydroxyl radicals at a plasma-liquid interface *Plasma Sources Sci. Technol.* **27** 115013

[22] Myers B, Dogariu A, Beeler B, Byrom L, Yatom S and Stapelmann K 2025 Imaging solvated oxygen atoms with a femtosecond laser *Nat. Commun.* (accepted)

[23] Marotta E and Paradisi C 2025 The importance of mechanistic studies in the development of cold plasma-based degradation of persistent organic pollutants in water *Curr, Opin. Green Sustain. Chem.* **52** 100999

[24] Costanza J, Arshadi M, Abriola L M and Pennell K D 2019 Accumulation of pfoa and pfos at the air-water interface *Environ. Sci. Technol. Lett.* **6** 487–91

[25] Fennell B D, Mezyk S P and McKay G 2022 Critical review of uv-advanced reduction processes for the treatment of chemical contaminants in water *ACS Environ. Au* **2** 178–205

[26] Jing C, Zhang P Y and Jian L 2007 Photodegradation of perfluorooctanoic acid by 185 nm vacuum ultraviolet light *J. Environ. Sci.* **19** 387–90

[27] Xin X, Kim J, Ashley D C and Huang C-H 2023 Degradation and defluorination of per-and polyfluoroalkyl substances by direct photolysis at 222 nm *ACS Es&t Water* **3** 2776–85

[28] Nzeribe B N, Crimi M, Mededovic Thagard S and Holsen T M 2019 Physico-chemical processes for the treatment of per-and polyfluoroalkyl substances (pfas): A review *Crit. Rev. Environ. Sci. Technol.* **49** 866–915

[29] Vasilev M, Conlon P, Bohl D and Thagard S M 2022 The effect of discharge frequency of a gas-liquid plasma reactor on bulk liquid transport and removal of organic contaminants *Plasma Chem. Plasma Process.* **42** 759–83

[30] Oinuma G, Nayak G, Du Y and Bruggeman P J 2020 Controlled plasma-droplet interactions: a quantitative study of oh transfer in plasma-liquid interaction *Plasma Sources Sci. Technol.* **29** 095002

[31] Delgado H E, Brown G H, Bartels D M, Rumbach P and Go D B 2021 The scaling of kinetic and transport behaviors in the solution-phase chemistry of a plasma-liquid interface *J. Appl. Phys.* **129**

[32] Srivastava T, Chaudhuri S, Rich C C, Schatz G C, Frontiera R R and Bruggeman P 2024 Probing time-resolved plasma-driven solution electrochemistry in a falling liquid film plasma reactor: Identification of HO_2^- as a plasma-derived reducing agent *J. Chem. Phys.* **160** 094201

[33] Peyres S M, Wang J, Hollyfield D W, Abuyazid N H, Sankaran R M and Üner N B 2024 Application of a film model to mass transfer and chemical reaction at a plasma-liquid interface *J. Electrochem. Soc.* **171** 113502
[34] Nau-Hix C, Holsen T M and Mededovic Thagard S 2022 Optimization of a gas-liquid plasma reactor for water treatment applications: Design guidelines and electrical circuit considerations *Plasma Process. Polym.* **19** 2200036
[35] Vasilev M, Stephens R, Muller M and Mededovic Thagard S 2024 Plasma bubble column reactor: a high throughput reactor design for water treatment *Plasma Chem. Plasma Process.* **44** 1773–96
[36] Thagard S M, Stratton G R, Dai F, Bellona C L, Holsen T M, Bohl D G, Paek E and Dickenson E R V 2016 Plasma-based water treatment: development of a general mechanistic model to estimate the treatability of different types of contaminants *J. Phys. D: Appl. Phys.* **50** 014003
[37] Ganzallo F J, Su X, Yatom S, Vasilev M and Thagard S M 2023 Characterization and treatment performance of an atmospheric pressure plasma jet-operated spinning disc reactor for the treatment of rhodamine B dye *Plasma Chem. Plasma Process.* **43** 1685–708
[38] Thagard S M 2025 Reactor design in plasma-liquid systems for wastewater treatment *Curr. Opin. Green Sustain Chem.* **53** 101023
[39] Nippatlapalli N and Locke B R 2022 Integrating nonthermal plasma with conventional technology for water pollution treatment *Int. J. Plasma Environ. Sci. Technol.* **16** e02006
[40] Kumar A, Škoro N, Gernjak W and Puač N 2021 Cold atmospheric plasma technology for removal of organic micropollutants from wastewater—a review *Eur. Phys.* J **75** 283
[41] Jenks L R, Bulusu R K M, Gallan R, Wandell R, Tang Y and Locke B R 2025 Degradation of low-concentration perfluoroalkyl substances with a nonthermal gas-liquid plasma reactor *Plasma Process. Polym.* **22** 2400219
[42] Nau-Hix C, Multari N, Singh R K, Richardson S, Kulkarni P, Anderson R H, Holsen T M and Mededovic Thagard S 2021 Field demonstration of a pilot-scale plasma reactor for the rapid removal of poly-and perfluoroalkyl substances in groundwater *ACS ES&T Water* **1** 680–7
[43] Metelmann H R, Von Woedtke T and Weltmann K D 2018 *Comprehensive Clinical Plasma Medicine: Cold Physical Plasma for Medical Application* (Springer)
[44] Metelmann H R, von Woedtke T, Weltmann K D and Emmert S 2022 *Textbook of Good Clinical Practice in Cold Plasma Therapy* (Springer)
[45] Bekeschus S and von Woedtke T 2024 *Redox Biology in Plasma Medicine* (CRC Press)
[46] Magureanu M, Bilea F, Bradu C and Hong D 2021 A review on non-thermal plasma treatment of water contaminated with antibiotics *J. Hazardous Mater.* **417** 125481
[47] Allabakshi S M, Srikar P S N S R, Gangwar R K and Maliyekkal S M 2021 Application of plasma-assisted advanced oxidation processes for removal of emerging contaminants in water *New Trends in Emerging Environmental Contaminants* (Springer) pp 333–70
[48] Schneider M, Rataj R, BláhaM L and Kolb J F 2023 Experimental review of different plasma technologies for the degradation of cylindrospermopsin as model water pollutant *Chem. Eng. J.* **451** 138984
[49] Banaschik R, Jablonowski H, Bednarski P J and Kolb J F 2018 Degradation and intermediates of diclofenac as instructive example for decomposition of recalcitrant pharmaceuticals by hydroxyl radicals generated with pulsed corona plasma in water *J. Hazardous Mater.* **342** 651–60
[50] Takeuchi N and Yasuoka K 2020 Review of plasma-based water treatment technologies for the decomposition of persistent organic compounds *Jpn. J. Appl. Phys.* **60** SA0801

www.ingramcontent.com/pod-product-compliance
Ingram Content Group UK Ltd.
Pitfield, Milton Keynes, MK11 3LW, UK
UKHW051924100926
8556IPUK00021B/1355